DeepAesthetics

Thought in the Act

A series edited by Erin Manning and Brian Massumi

DeepAesthetics

Computational Experience in a Time
of Machine Learning

ANNA MUNSTER

DUKE UNIVERSITY PRESS
Durham and London
2025

© 2025 DUKE UNIVERSITY PRESS
All rights reserved
Printed in the United States of America on acid-free paper ∞
Project Editor: Bird Williams
Designed by A. Mattson Gallagher
Typeset in Minion Pro, Source Sans 3, and Adobe Arabic
by Copperline Book Services.

Library of Congress Cataloging-in-Publication Data
Names: Munster, Anna, author.
Title: DeepAesthetics : computational experience in a time of machine learning / Anna Munster.
Other titles: Deep Aesthetics : computational experience in a time of machine learning | Thought in the act.
Description: Durham : Duke University Press, 2025. | Series: Thought in the act | Includes bibliographical references and index.
Identifiers: LCCN 2024039427 (print)
LCCN 2024039428 (ebook)
ISBN 9781478031543 (paperback)
ISBN 9781478028338 (hardcover)
ISBN 9781478060529 (ebook)
Subjects: LCSH: Deep learning (Machine learning) | Machine learning—Social aspects. | Artificial intelligence—Social aspects. | Technology—Social aspects. | Knowledge, Sociology of.
Classification: LCC Q325.73 . M86 2025 (print)
LCC Q325.73 (ebook)
DDC 006.3/1—dc23/eng/20241226
LC record available at https://lccn.loc.gov/2024039427
LC ebook record available at https://lccn.loc.gov/2024039428

Cover art: Anna Ridler, *Mosaic Virus*, 2019. Still from 3-screen GAN video installation. Image courtesy of the artist.

For all my collaborators, human and more than.

CONTENTS

	Introduction: Deep Machines and Surfaces of Experience	1
1	**Heteropoietic Computation** Category Mistakes and Fails as Generators of Novel Sensibilities	40
2	**The Color of Statistics** Race as Statistical (In)visuality	78
3	**Could AI Become Neurodivergent?** Disfluent Conversations with Natural Language Processors	114
4	**Machines Unlearning** Toward an Allagmatic Arts of AI	147

Postscript On Models of Control and (Their) Modulation	177
Acknowledgments	183
Notes	185
References	201
Index	219

Introduction:
Deep Machines and Surfaces of Experience

An Experience of Computation

Cryptozoology may seem as far-flung from data science as paleontology or alchemy. But in April 2022, a Swedish musician and artist working on a then obscure generative art strategy burst an artificially intelligent "cryptid," or mythically existent creature, onto the text-to-image artificial intelligence (AI) creative scene. Cryptids are animals that populate folklore, subcultures, parapsychology, and, increasingly, the internet. They exist *in the wild*, in wild places; they are creatures mainstream science refuses to verify. After several months of experimenting, Steph Maj Swanson introduced her proliferating images of something she called "an emergent phenomenon that arises in certain AI image synthesis models" (Swanson 2022a), via her "Supercomposite" Twitter account. She named the woman who

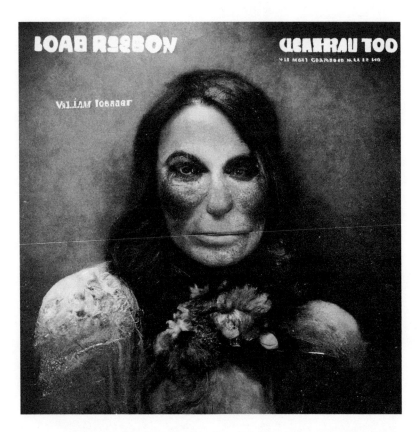

I.1 *untitled progenitor Loab 2*, one of the first two images of Loab posted by Steph Maj Swanson, using the artist name Supercomposite, on her Twitter account, September 7, 2022. Image courtesy of Steph Maj Swanson.

regenerated, seeded, and spread across these images "Loab." Within two days, however, Loab was proclaimed the first "AI-cryptid" by eBaum's World (Zachnading 2022), an online meme-gathering and entertainment site; and within a week, the powerhouse British fashion and culture magazine *Dazed* had likewise confirmed her cryptid arrival (Waite 2022).

Not only had Loab become an emergent AI phenomenon; she had also emphatically implicated the unverifiable creatures of cryptozoology in (data) science's latest shiny enterprise of large-scale language models with generative capabilities.

Driven by the machine learning (ML) capacities and infrastructure of such models, AI had already been heading in one or several of such trajectories for some time: deepfakes standing in for the media presence of actors and presidents; AI influencers on Instagram and TikTok with millions of followers; lifelike computationally synthesized portraits of nonexistent human faces. But the efforts of artists, developers, and data researchers in crafting text "prompts" and finding new ways to conjure and intervene in the ubiquitous culture of statistical computation garner less notoriety and commentary. Loab was not hiding in the "black box," however, waiting to crawl out from under a wild AI place; she was prompted. While Swanson notes the emergent properties of AI, she nonetheless introduced the Loab image series and genealogy via a lengthy Twitter thread on process and AI text-to-image generation.[1] Swanson had begun with an experiment in scripting an image prompt—the action of entering text into a field in an interface to generate an image by an AI/model. Models offering these interfaces, such as Stable Diffusion, Midjourney, and DALL-E, became hugely popular throughout 2022 and were accessed via web browser and dedicated apps or run as stand-alone models on desktop computers. They are part of a suite of deep learning models that engage, augment, and extend the computational architecture of large language models (LLMs), which I begin to unpack in chapter 1 and take up in more detail in chapter 3. As with much AI, technical developments in model architecture, training, and operativity are both incremental and swift. Throughout this book, I offer examples of algorithms, models, and practices that may seem specific to a particular historical and technical development or may appear no longer to be widely in use. Yet my selection of such examples often rests on their ongoing if unremarked deployment by machine learning or, in the case of Loab, because the operation performed on or with them becomes a way of probing the specificity of machine learning computational experience.

Swanson's experiment involved doing something slightly different with text-to-image models: "negative prompting," or writing a script for the opposite of what the prompt will sample from the image space. In negative prompting, the model samples images the furthest statistical distance away from their match to the text written in the prompt itself, that is, furthest along a distribution of text and image-matched data on which the model has been trained. In the case of the prompt that began the generation of Loab, Swanson tried for a statistical negative of the text "Marlon Brando," entering the script "Brando:–1." This returned an oddly banal imagemash of a nonexistent company logo, "DIGITA PNTICS," set against a seemingly hand-drawn schematic of a skyline, with text deformations typical of the model's inability to graphically render text. Swanson became curious about a double negative prompt that might then send the model into re-generating Marlon Brando's image when prompted by the negative of its negative (prompt): "DIGITA PNTICS skyline logo:: –1." Instead Loab appeared, affirming her imagistic rise as the model's response to the negative of Marlon Brando–esqueness.

Although Loab, the AI-generated woman, has been called a cryptid, a "demon" (Ryan 2022), and a queer icon (Levanier 2022), Loab, the AI images generated by text-to-image ML, can be understood a little differently as what I will call "process probes." These are techniques that Swanson developed to sift and drift through the latent spaces of AI's generative imaging. As I explain in chapter 1, latent space is always specific to an AI's or model's distribution of the data points it reorganizes when training on an original dataset. As a model trains, it organizes and distributes its input data or, in some generative image models, the randomized noise data it is fed via a particular algorithm or function. These input data are clustered into groups or features by the model's operations. As data are reorganized, a distribution takes shape that becomes part of the learning the model gains about the data, a distribution in which the features are contoured by relations of close or distant resemblance and proximity. When Swanson artfully scripted her prompts—with Marlon Brando–esque, then negative Brando-esque, then the negative negative of Brando-esqueness—she prompted the model to sample across its trained distribution of data proximities and distances (or resemblances and dissimilarities). It is this overall configuration that contours text and image-paired data into what is called the latent space of a generative text-to-image model. Swanson developed and stumbled on Loabness out of Loab's "latent" potential phenomenality through her artful and curious probing of the blind, relational, and potentially wild spaces of machine learning-driven AI. Swanson both stumbled across and developed a *co-loab-oration* with the model, finding something odd, lurking, but barely there as "an emergent island in the latent space that we don't know how to locate with text queries" (Swanson 2022b). Loab was the output of a process that artfully explored the unknown unknowns of what is otherwise touted as predictive computational experience.

Experiencing DeepAesthetics

More than an artwork, more than a collaboration between AI and human, Loab affords us a particular mode of *experiencing* computation. If indeed she-they/Loab-Swanson probes the processes and relations that make up the physically nonexistent yet real statistical terrain of latent space, then she and they occupy and help generate a radically different idea of and encounter with computational experience. This has nothing to do with the identity of the woman or artist behind the Loab image. This has to do with conceiving experience differently or rather *differentially*. I take up William James's philosophy of experience in the context of probing ML and its strange emergent Loab-like phenomena throughout this book because James's concept of experience can address machine *learning* as processes that change, and as processes for experiencing computation changing. For James, and process thinking more broadly, change unfolds; *it becomes*, via processes of continuity or conjunction, and discontinuity or disjunction. This offers us experience based not on identities or positions such as the human, computer, or AI but rather on undulating fields shot through with continuously changing relations. James furnishes a conception of experience philosophically placed to one side of the "lived experience" from the phenomenological tradition and, different again, from explorations through feminist, race, and queer politics and theory, still preoccupied with all too *human* embodiments and trailing under the long tail of identity politics. Rather than experience being "purified" and reified to some primary "ground truth," body, or position, James's "pure experience" welcomes all, any, and every experience. Crucially, pure experience is not made by or filled with things or places such as "subject" and "object" but is generated through relations and processes, which James terms "co-ordinate phenomena" organizing its space-times (1977, 199). So too do relations organize the space-time of *computational* experience and its weird yet powerfully generative topologies. And while Loab has been described as a ghostly haunting of AI and a "creepypasta" or internet horror figure (Ryan 2022), the Loab imagescape registers something real: the statistical reconfiguration of experience by ML computing. As computation has increasingly been inflected by ML, our cultural, computational, and medial outputs as online images, generative artworks, and text corpuses—indeed, all and any data—have been modeled into maximum and minimum clusters of proximities that simultaneously butt up against one another as continuous regions or disperse away from one another via discontinuous edges and outliers.

Now that ML is so pervasive a form of enacting computational processes, contemporary experience has become littered with all manner of imperceptible statistical relations. Many of these take place between humans and computers, and many others among computers or computational elements alone. Collectively and differentially, these multiply scaled, differentiating relations change the stuff of experience, change all those living and technical elements experiencing, and generate new relations that unfold into many ways for making futures. As the hegemonic form of computing today, ML encompasses a diverse terrain of systems theory, practices, and applications that build and modulate computational models in relation to inputs or data (Alpaydin 2016, 17).[2] The capacity of the computational model or AI to change in relation to (changing) inputs is what is understood by data science as "learning." These modulatory and adaptive systems underpin everyday (human) experiences such as online shopping, streaming music, and airport security by using, respectively, recommenders, collaborative filtering, and biometric recognition. And they are now organizing and contouring swathes of individual and collective engagements and encounters. Loab skips across many ordinary and extraordinary aspects of how ML has reconfigured computational experience—from the ways in which entering a text prompt can now generate slabs of generic text reportage and writing being used by students to write essays, to the ways in which text can artfully be rescripted to create uncanny imagery. Loab, then, sensibly registers the processes that together underpin and generate ML experience.

In a definition of ML often quoted by data scientists, *experience* is a key term in judging whether a computational system qualifies as one that learns: "A computer program is said to learn from experience E with respect to some class of tasks T and performance measure P, if its performance at tasks in T, as measured by P, improves with experience 'E'" (Mitchell 1997, 2). Here experience, or E, appears twice in the proposition, but importantly, its recurrence suggests change. E is first a defined phenomenon—it could be a measurable state of a set of inputs, for example—on which a series of algorithms performing tasks (T) have run (P). But it can also become the change taking place—the "improving," for example—from which a measure of the tasks' improved performance is taken. Tellingly, that measure of improvement is ascertained by running an ML program or model many times over the data while it is in its primary learning phase or training. Here, E varies as the model's learning attempts to recognize a structure or pattern in the data. Experience, then, for ML, is simultaneously quantifiable as a state of measurable change *and* the ongoing process of learning that variably qualifies what that change is to be over time. The final improvement, or what the model has learned, is therefore really a coalescence of many processes of modulation differing from and conjoining with one another. As changes or learning occurs, this modulation—which we will come to know as the model's operativity—*qualifies* the entire ML ensemble of data and functions. Machine learning experience is an ellipsis of the two experiences—of what has occurred and what is occurrent. However, ML is typically researched, reported on, and implemented as a "learning problem" to be solved (Mitchell 1997, 3). Rather than the dynamic ellipsis of past and present, E is more often than not reduced to a quantifiable ratio of improvement measure against a set of inputs on which a model trains, such as a database of faces for training a facial recognition AI. The ongoing learning then becomes E as the "measure of improvement" against the E of the trained model. We lose the dynamic set up by the vectors of the two ES continually traversing the occurred and occurrent, and traversing the temporalities of a backward and forward. We also lose what this ellipsis of present and past, past and future, suggests about a double process performed by ML. This doubling involves both a reaching across and a contraction of quantity with quality. Such slippages, extensions, and mergers from quantity to quality and back are at the relational heart of the computational experience of ML.

To take up these pulsations rhythmically, I want to propose that deepaesthetics, a concept I use to think computational experience in this book, is likewise occurring via different rhythms of expansion and contraction. Deepaesthetics offers us both conjunction and disjunction, gluing together two worlds that do not seem to be of concern to each other: deep learning, the subfield of machine learning that uses neural network architectures; and a branch of philosophy traditionally concerned with how valuations of formal or sensory qualities come to be made. I am interested in how that contraction actively sticks together through the operations performed by ML computational assemblages, through the creativity attributed to ML-driven AI, and via actual artworks that stage encounters with ML. I am also interested in how it splits apart, rupturing as the forces across computational and human experience difference each other.

Aesthetics is, of course, much more than a branch of philosophy; even philosophically in the Western canon, its history is complex and ambiguous. Since the eighteenth century, philosophical debate has oscillated between the different positions taken up by, on the one hand, Immanuel Kant's categorization of aesthetic judgment in his 1770 *Critique of Judgment*, which ultimately grounded aesthetic value in a disinterested appraisal of perceived, sensory phenomena (1987, 44), and, on the other, Alexander Baumgarten's *Aesthetica* of 1758, in which he considered all sensory experience to be aesthetic. The contemporary fallout of this legacy for an aesthetic consideration of computation has been, largely, to fall on either side of a formalist or sensorial approach, although in the ensuing aesthetic debates, neither formalism nor sensorialism maps back neatly onto Kant and Baumgarten. Aspects of formalism characterize the work of Beatrice Fazi's (2018a) computational aesthetics in her argument that the digital must be taken as a formally autonomous realm, whereas, for example, Mark Hansen's approach has been to continue to understand computation as a "phenomenotechnics," a kind of entanglement of the technical with lived, sensorial experience (2021).

However, computational experience is entangled with modes of sensing that are not only beyond human sensing but, as Matthew Fuller and Eyal Weizman argue, beyond *perception*: leaves that become sensitive to herbicides and whose sensitivity might be measured via biosensors, sensors encoded to detect respiration and moisture rates in greenhouses, sensing undertaken by computer vision models that detect variability in the forest canopy (2021, 33–50). This makes both a formalist and an embodied sensorialist aesthetics tricky. Fuller and Weizman argue that sensing—whether computationally enabled, augmented, or occurring outside of computation—involves events in which all kinds of surfaces register and inscribe their contact with one another, events that can be ordinary and everyday as well as technically refined and deliberate. They call this panoply of sensing events coursing throughout the world "aesthetic," which entails the aesthetic as just that ubiquitous domain of all sensing relations. Their argument concerning aesthetics as the potential for the registration and inscription of sensing on any and by every surface whatsoever resonates with my Jamesian approach to (computational) experience throughout this book. As I explain a little later in this introduction, experience understood via process philosophy comprises largely any and every relation in its/their process and quality of relating and registering these relations.

But, as Fuller and Weizman explain, registration and inscription events *make sense* in different ways, since even ubiquitous relations do not register evenly or with the same qualities for all entities or surfaces in relation. There is always a making sense accompanying sensing—what they refer to as "sense-making"—that involves varying "cultures of sensing" (52). Cultures should not be understood as only comprising human subjects who make sense of objects. Rather, cultures of sensing work via layers and accumulations of sensing that accrete materially, institutionally, and perspectivally under situated and differing histories and assemblages. And these are invoked to *make sense* of sensing.[3] The strata of such formations are never frictionless but involve tensions of scale, perspective, materiality, and power. Aesthetics, for Fuller and Weizman, is this bringing into relation of sensing events with cultures or formations of making sense and can itself involve tensions.

In this book, I take Jamesian "experience" as the broader term with which to begin, since it is always already about the reality of relations in which surfaces and strata of computational, sentient, or any matter are eventfully in contact, registering and prehending each other. But, like Fuller and Weizman's notion of aesthetics, this does not mean that experience is self-similar in its relationality or registration. In this book, I understand aesthetics as modes of individuating this broader field(s) of experience, so that experience comes to make sense via singular sensibilities, whose accretions and formations may well be riven with tensions. Machine learning engenders a specific individuation of computational experience in which we are both asked to encounter and bound to insensible and microperceptible forms of nonlinear and continuously modulating statistical function and calculation; this is its aesthetic condition. This poses a problem for sensing and sensibility: How can we perceptually register and even account for what occurs computationally at scales, durations, and dimensions that are nonhuman and, frequently, imperceptible?

In the now famous deep learning research that accompanied Google's Inception model (see Szegedy et al. 2015), hallucinatory synthetic images of dogs, birds, bananas, and more seemed to have emerged via an imperceptible process from a starting point of visual noise. To make this process explainable, Google's researchers developed an entire visual online site stepping through the movements from noise to recognizable animal/object in an image.[4] However, an entire aesthetic individuation is mobilized around this explanation of the functioning of Inception. This individuation relies on a representationalist paradigm of (visual) perception widely deployed throughout deep learning models, which I draw attention to in chapter 1. Here the desired representation of an object seems to emerge via continuous steps out of an initial flux: "Start with an image full of random noise, then gradually tweak the image towards what the neural net considers a banana" (Mordvintsev, Olah, and Tyka 2015). Indeed, the visual layout accompanying explanations of how features work in neural networks often reinforces this steady building up of a representation. However, imperceptible ML processes are operating and registering at the same time, cutting into the continuity of this aesthetic of representationalism; as the Google researchers admit, "By itself, that doesn't work very well, but it does if we impose a prior constraint that the image should have similar statistics to natural images." What cannot be visually represented in the stepping through of features, then, is just that statistical "prior constraint" learned and transduced into a statistical weighting from a different distribution and dataset of "natural images." Yet both constraints and distributions are key operations of deep learning models and are crucial to how their sensibility registers.

My use of a term such as *deepaesthetics* aims to work productively with this problem of how to sense or register what is occurring imperceptibly, whether that involves operations, statistical techniques, or the qualities of relations in ML's computational experience. Across the many sites of ML practices, techniques, and operations in this book, I focus on just those *continuities and discontinuities* that characterize the registration of ML and its processes, both by its own surfaces and by surfaces of human sensing. My wager will be that by bringing careful, granular attention to ML's processuality—that domain of computational experience registering yet often beyond perception—we can scope out what is singular about the sensibility engendered by its aesthetic individuation(s).

The Depth of Deep Learning

Machine learning architectures come in many forms, but a frequently used one is the deep learning neural network, in which complex computational representations of empirical phenomena are developed through "layers" of numerical values that then represent low-level up to higher-level features capable of synthesizing new data representations. In what has become a standard text on deep learning for ML research, Ian Goodfellow, Yoshua Bengio, and Aaron Courville define the "deep" of deep learning as the method and architecture for computationally resolving the problem of representation: "Deep learning enables the computer to build complex concepts out of simpler concepts" (2016, 5). From low-level features, increasing functions and processes add, subtract, and multiply in linear and nonlinear ways to increase the complexity of the representations or syntheses. Eventually, after x epochs of training, or synthesis, this results in new data representations/outputs that correlate with the inputs or are sufficiently synthesized to efficiently accomplish a given task. Examples might include a neural network learning the representation of any handwritten letters in an alphabet it observes after training on a database of diverse-enough alphabetic letters, or a network being capable of synthesizing realistic-looking photographs of cats in a range of positions after training on many still video frames of cats in indoor and outdoor settings. Here the deep aspect of the neural network is measured by the number of layers of parameters (constrained sets of numerical relations) through which data inputs—handwritten letters, images, text, music, and so on, transduced into numerical values—pass. To complex representations that are perceived by human sensing, "a deep learning system can represent the concept of an image of a person by combining simpler concepts, such as corners and contours, which are in turn defined in terms of edges" (Goodfellow, Bengio, and Courville 2016, 5). Part of this project for probing and registering a deepaesthetics involves finding these kinds of *discontinuities* that are lodged in the operativity and rhetoric of deep learning and pose as smooth continuities across computational and human logics and modes of perception.

Layers also mark a kind of gateway or threshold for these sets of values to pass to or from the next layer/set and then across the neural network. Throughout the book, I explore some of the commonly used technical terms that interweave ML research and practice via pop-up definitional boxes and accompanying images and diagrams. These will facilitate our encounters with the technics of ML, giving us provisional means to navigate its techniques and operations. The purpose of these is not to comprehensively define all technical terms throughout the book. Instead, I touch on technical elements that recur and also often work ambiguously or opaquely within data science. These pop-ups are, then, terms to watch out for, terms that return, and terms that trouble the field.[5]

Depth in neural network architectures also refers to the multiplication of layers, which occurs in larger and more complex models. Learning is then understood to occur as the network successively "discovers" across these many layers of features and patterns detected in or generated by inputs (LeCun, Bengio, and Hinton 2015, 436). These layers are often referred to as "hidden" in the sense that both their location and knowledge of their exact functionality in the model may not be precisely discernible. The deeper the network or the more layers it has, the better able—the claim is made—it can train for fine-grained features. Once trained, the model will have learned how to accurately detect and predict the features of unknown data inputs. In some generative networks, deep models can create new synthetic instances of data; we are familiar with these through deepfakes and with images and videos rendered in response to text prompts by post-GPT-3 (Generalized Processing Training) large-scale language models.

Deep learning networks can deploy millions of parameters at successive layers that adjust during a model's training. Here I am describing only some of the general characteristics of deep neural networks—inputs, layers, parameters, learning, output, and prediction—to grasp the ways in which data science conceives depth as a horizontal stacking of connected layers, which

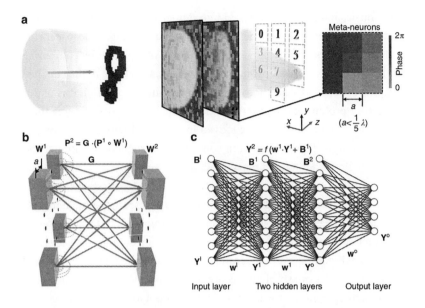

I.2 The three schematics represent the layers of the neural network architecture: first, projected two-dimensional geometric surfaces in *a*; then as a snapshot of dynamic calculative processes producing a network topology across the orange lines in *b*; and finally named as layers in which groups of values are being calculated and passed forward from input to output at *c*. From "Meta-Neural-Network for Real-Time and Passive Deep-Learning-Based Object Recognition" (Weng et al. 2020).

are perceptually inaccessible for human registration. When we look at common figures that schematize deep neural networks (fig. I.2), we immediately see that depth is imaged in avolumetric terms, if we understand volume to be a Euclidean geometrical arrangement of measurement between *x*, *y*, and *z* axes. Where layers might suggest the potential for depth to accumulate via a vertical stacking of each on top of the next, the schematics of deep learning's *operations*—its processes—perform otherwise via lateral, relaying, and recursive movements. The diagrams that describe deep learning's models must be understood, then, as inhabiting a different kind of space: a topological one of relation and process.

For data science, the "deep" in deep learning is also a quantitative problem, which arises out of these more-than-volumetric spaces of data-operation topologies. Much of the data being crunched through deep models has high variability; every pixel of an image data input, for example,

Pop-Up Definition: Layers in Neural Networks
Layers are surfaces composed through the topology of a neural network's numerical and functional (algorithmic) relations. They are often represented in ML diagrams of neural network architectures as geometrically projected two-dimensional surfaces. In actuality, they comprise calculations, extracted sets of values, and vectors produced by functions of a particular neural network.
The definition of *layers* varies in machine learning and often conflates concepts such as surface, gateway, feature extraction, and threshold. At its most basic, a layer is simply a set of data points either extracted from data inputs or computationally synthesized, which have been constrained by certain parameters (filters, weights, and so on). These constraints extract or configure certain values from the data or perform a synthesis according to a set of values, and the result of either operation is known as a "feature." These new sets of features (or new sets of numerical values) are passed on to the next layer for further extraction or synthesis.

has contrast, hue, luminosity, and saturation features, among many data points, and holds these in relation to all other data points within the same image. Together this multiplies data's relationality both intensively and extensively and is known within data science as "high dimensionality." The relationality encountered here is likewise avolumetric and can only be spatialized *n-dimensionally* (any number of dimensions), often via what we recognize as topological network diagrams (fig. I.3a, fig. I.3b). To discover across *n*-dimensional features of the data only those patterns of relevance for a specific task, ML must manage swarms of both irrelevant and less relevant features. Its management techniques involve statistical operations that compress the volume of the data's high dimensionality. This suggests that a certain qualitative flattening must be implemented for the efficient functioning of the model, which, as we will see in chapter 2, may be enacted via statistical functions. These functions or algorithms are often performed before the operations of the *deep* learning neural network architecture. Data, then, may be subjected to several other ML algorithms and processes before becoming the inputs on which an AI deep neural model learns across its layers. Such algorithms and processes seem much less spectacular than the various deep networks capturing scientific and public attention in the last decade, such as DeepMind's AlphaGo, DeepDream, DALL-E, and deepfakes.

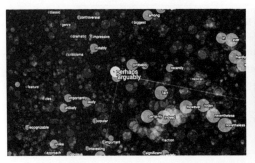 

I.3 Two different visualizations of high- or *n*-dimensional space of text data: the left panel showing global proximate and distant clustering of data points to represent similarity and difference via proximity and distance; the right panel showing detailed connectivity of words in the text to each other. Note that these visualizations are manipulable in three-dimensional computer graphic space, which also shifts the "view" onto them.

Pop-Up Definition: Dimensionality Reduction
All data have attributes or "dimensions." These might include age, sex, gender, and so on, for demographic data, and pixel color, brightness, and saturation, and so on, for digital images. Rich data such as images have many attributes for each pixel, and so their data are called "high-dimensional." When algorithms and machine learning models attempt to locate features in high-dimensional data, they may be slowed down, become inefficient, or "distracted" by unwanted attributes. Data science has historically drawn on and remodeled statistical techniques for reducing the dimensionality of data.
Dimensionality reduction algorithmically removes attributes or dimensions from a dataset that are not seen to be intrinsic to the patterns, features, or tasks being trained for, discovered, or recognized in the data. In data science, dimensionality reduction is understood to quantitatively reduce data but not to change its overall qualitative characteristics.

And yet the stacked horizontality of layers and dimensionality reduction are two key aspects of the strange topology of surface-generated depth through which machine learning–driven AI works. I will often need to navigate such functions and their sociotechnicalities alongside the deep learning models that have attracted the most attention. For this reason, I take *machine* learning, which encompasses an array of techniques, operations, and processes of statistical computation, as the larger domain of computational experience in which deep learning and generative AI are embedded.

The routinely performed processes of dimensionality reduction in ML are, however, not simply quantitative operations. They simultaneously organize the relationality of data according to vectors qualified by similarity, difference, and their interrelations. Data science attempts to measure or quantify such vectorization through functions that calculate maximum and minimum distribution of sameness or difference. Nonetheless, these functions qualify the volumes of data by shaping them into differential clusters, and this interpolates a more-than-quantitative register in/with the data. At the very moment that data come to be quantitively operated on by statistical methods such as dimensionality reduction, the data are also being respatialized and reconfigured with hidden potential for machine-discoverable pattern, recognition, and classification. Patterns or recurring motifs and classification or discrete separation would not be possible without the vectorial shaping of data, a shaping that is qualitatively immanent to quantized organizations of data. As Adrian Mackenzie puts it, when discussing the ways in which the vectorization of prostate tumor data can arrange and align features within a dataset: "The question of relation between multiple variables and . . . predicted levels . . . suggests the existence of a hidden, occluded, or internal space that cannot be seen in a data table and cannot be brought to light even in the more complex geometry of a plot. This volume contains the locus of multiple relations, a locus inhering in a higher dimensional space" (2017, 63). Even while such operations on and with data are quantitative, they also change the configuration of the data's intensive relations as the model learns: "Deep neural networks operate by transforming topology, gradually simplifying topologically entangled data in the input space until it becomes linearly separable in the output space" (Naitzat et al. 2020, 35). Depth, then, reemerges as what resides within both model and data yet cannot be seen or calculated exactly. These deep spaces emerge as data's dimensions are reduced, but they also signal a computational register that cannot be fully circumscribed by performing quantizing calculations. The "deep" in deep learning endures just beyond the measurable. This suggests a conundrum characterizing ML experience insofar as neither humans nor deep learning models seem to possess the capacities to engage those very qualities orienting and characterizing AI's operativity. Rather than quantifying, cognizing, or visualizing computational experience in a time of ML, I will propose different modes, levels, and registers for experiencing (its) experience.

With its myriad operations, novel spaces, and dynamic transductions of quantitative phenomena to qualitative events, contemporary computational experience lends itself to being thought and felt processually. Throughout the book, I will have recourse to concepts from process philosophy to understand machine-learning-based AI. Rather than focusing on linear functions seamlessly chaining inputs to outputs or to nonlinear algorithms equally striving for error-free prediction, I will focus on the recursive and modulating functioning of ML, whose processes, while calculable, are not in themselves necessarily determinable.[6] An often-heard proclamation in data science is that deep learning, in particular, is a black box: it functions, but we don't know what goes on inside (see, e.g., Castelvecchi 2016). Instead of pursuing what is determinable in the black box—which, in deep learning, has become a research domain in its own right—I will suggest that more might be gained by thinking ML experience as and through processual operations. A significant benefit of doing so is that it allows us to hold together the many tensions and knots crossing the quantitative and qualitative, the calculable and the indeterminate, the discrete and continuous, *as the very stuff of experience* in a time of ML.

A Radically Empirical Experience for and of ML

I am not the first to propose that the recursive processes of ML-driven computation lie at the core of its contemporary operativity. The updating of both data and ML systems is also commented on by Taina Bucher, who draws attention to the ways in which algorithms are in a constant process of becoming as technical, social, and ultimately governing forces and events (2018, 28). Closer to the approach I offer here is Luciana Parisi's project for affirming the incomputable as those indeterminate quantities of data produced through the recursive and nonlinear operations of computational modeling (Parisi and Dixon-Román 2020; Parisi 2013). Parisi has argued that the shift to many-layered deep learning formations of AI sets up nonlinear recursions across the model as it runs. These recursions generate a kind of extra-dimension of data from which the model itself adapts and learns but for which it can never fully calculate or compute, since its ongoing operativity maintains this very excessive generativity: "This wall of incompressible data instead overruns the program and this neutralizes or reveals the incompleteness of the axioms on which the program was based in the first place" (Parisi and Dixon-Román 2020, 57). A margin immanent to the neural architecture of AI exposes itself, asserting a gap between the model's claims to prediction and determination and its engendering of an excess from its autopoiesis.

However, for Parisi, and likewise for Fazi (2018a), recursive and contingent computation produces indeterminacy by purely quantitative and axiomatic means. For Fazi, this occurs because computing is based on Alan Turing's universalizing axiom—the ongoing operation of a machine's determination to execute either state A or state B. Yet in its inexhaustible continuity, computing must necessarily encounter numbers such as infinity, which are *incomputable*: "With Turing's incomputability we are witness to something especially surprising: it is the mechanical rule itself that, in its own operations of discretisation, generates the inexhaustibility of computational processing" (Fazi 2018a, 124). Fazi sees the simple digital process of deciding A or B, 0 or 1, a binary and determinate action, as the potential for computing to set off on a path toward indeterminacy. Crucially, for these theoretical approaches to contingency, which pursue the quantitative and axiomatic, computation exists in a separate domain from empirical experience. The former is purely calculative, whereas the empirical is apportioned to the field of the sensible. And in the empirical realm, contingency or indeterminacy only arises through material or sensory conditions and circumstances.

But what if neither experience nor the empirical are primarily sensory, that is to say, sensory in the first instance? What if this division between the formal/axiomatic and the material is a secondary division of the dynamic ongoing reality of the world as it occurs in the making, including computational worlds? As we have already seen in ML, experience can be a measured phenomenon such as an input or output on which the model runs. But it is also—and crucially for an ontogenesis of *autonomous systems*—occurrent learning, or computation as the *differencing* generated as computing, the process, happens. If we think just of an AI model as a limited instance of computation, we locate that change or difference as the experience the model gains across its network by vectorially mapping the *relations* of inputs to outputs. We cannot reduce the function of this to any causal or linear mapping of data inputs to outputs, since most neural networks function via combinations of nonlinear relations such as pooling, back propagation,

INTRODUCTION   17

reinforcement learning, and so forth. Here "nonlinear" means that the outputs and inputs cannot be directly algorithmically mapped to each other.

What we can state is that any kind of ML understood as "a computer [that] is said to learn from experience E" involves *relations* of comparison, contrast, addition, subtraction, and multiplication in which both model and data configure and reconfigure through modulation, that is, the work of ongoing change/differences. And, more broadly, ML experience, as technical—and, as I will engage it in this book, as cultural, aesthetic, and social—must be taken more broadly as change occurring via the multiplicity of computational *relations* on the move. These encompass the model's algebraic relations but also those across data and its preprocessing via operations such as dimensionality reduction; the vectorization of data by a model; the differences produced by relations between neural networks in an assemblage of models, which are often used to accomplish complex tasks such as AlphaGo's chess wins; back propagation (used to efficiently calculate the multiple derivatives produced by the model computing the many variables of data inputs); optimization (which makes the model run with a reduced error rate); and a multitude of human intelligence tasks with which all the purely computational operations might also be entangled.

In this book, I unfold an approach that emphasizes the vectorial and *qualitative* operations performed by ML. These range from statistical functions such as principal component analysis (PCA), part of ML's array of algorithms, to the complex, dispersed, layered, and recursive architectures of multilayered neural networks. These qualitative operations are always exchanges between and across the quantitative (data) and axiomatic (algorithms or functions) and qualitative operations such as recursion, vectorization, and so on. Or, rather, we could say complex relations of sameness and difference traverse computational quanta, functions, and operations. It is this operative *relationality* that accounts for ML's contingent and nonpredictable modes of computation in which novel spaces and sensibilities form. And while these spaces and potentialities are insensible, this does not foreclose their registration as a sensibility specific to ML. This is *machine learning experience*: production and registration of a peculiar computational experience. And an ML sensibility can also be artfully conjured and encountered. It is part of the project of this book to signal where and how such encounters occur in the work of artists, cultural producers, and sometimes experimental data scientists interested in an alternate deepaesthetics.[7] Indeed, experiments with ML's relationality are already occurring, exploring its potential to open to the nonknown. In an experiment with the transla-

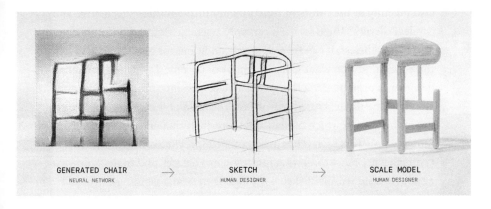

I.4 "Steps from Generated Image to Sketch to Physical Model." Philipp Schmitt and Steffen Weiss, *The Chair Project*, 2018. Image courtesy of Philipp Schmitt.

tion of synthesized images of chairs generated by a generative adversarial network (GAN) model, Philipp Schmitt and Steffen Weiss (2018) took the odd, dysfunctional deep-learning-generated images of blurry chairs to aid the human design of physical, albeit speculative, "chairs." Images generated by GANs, with their transmogrified snapshots of their model's learning of prominent features across a training dataset, have become readily identifiable as an aesthetic visual style of ML. But rather than veer either toward aesthetic realism—where the visual objective is to get the model to synthesize a realistic-looking chair via training on a dataset—or toward the "latent-space" style associated with GANs, Schmitt and Weiss's *The Chair Project* does something different.

The GAN-generated images become visual prompts that probe and produce a relationality across model and (human) designer: "The idea was to neither simply trace the generated images, nor to transform it into traditional pieces of furniture. Rather, we brought out the chairs we saw in the blurry images to help viewers see what we imagined. 'Seeing the chair' in an image is an imaginative and associative process. It pushes designers away from usual threads of thinking towards unusual ideas that they might not have had otherwise" (Schmitt and Weiss 2018, 2). The resulting physically crafted chairs are emergent realizations that embody the processes of back-and-forth prompting and probing across model and designer and across human perception and computational sensing. The chairs hint at a classic modernist design lineage bound to notions of "form follows function" while also pushing at the limits of functionality through their speculative and "useless" design. The GAN itself can only assemble features; it cannot see a chair. Nonetheless, it can furnish novel conditions for seeing to occur, and these can condition what seeing might be or become. Likewise, the human designer is prompted via the speculative constraints, probes, and parameters that the model blurrily furnishes. The artfulness of *The Chair Project* lies in the ways in which human and more-than-human forces collectively individuate across each other to co-compose previously nonknown chairs: chairs that are more than what is already in a dataset, and more than what a trained designer might sketch through their own imaginings.

The artful probing of ML's sensibility enables rich and generative encounters with computational experience that often question and sometimes even upturn a more predictive teleology. But to fully account for ML as a relational mode of computing, we need to think with process thinking. James's thinking has the advantage of valuing relations as the real stuff of experience, providing a deceptively simple definition of relations as "different degrees of intimacy" (1977, 196). By intimacy, James means a proximity of connection that produces varying degrees of transition in experience: "With, near, next, like, from, towards, against, because, for, through, my—these words designate types of conjunctive relation arranged in a roughly ascending order of intimacy and inclusiveness" (James 1912, 43). His emphasis on conjunction as a key form of transitions occurring revalues relations of continuity. Continuity is always generated qualitatively via relations that modulate things as they also change their milieu: "Not only is the *situation* different when the book is on the table, but the *book itself* is different as a book from what it was when it was off the table" (James 1977, 223). Continuity, then, varies and, as it does, modulates into different experiences of situations, things, and the entire ensemble of their relations. In the ascending order suggested by James, conjunction moves from "with" to "my," building from exterior bare relations to intimate subject-oriented perspectives, terminating with a human-subject experience of relation to "their" world. However, James is emphatic that no one connection or ordering runs through all experience (1977, 197). This then makes his conception of experience an open relationality, potentially made and individuated by all kinds of entities, including technical ones. As David Lapoujade puts it, "Pure experience is the set of anything that is in relation with something else" (2019, 13). The empirical is just that domain of any and every relation in its/their process of relating, but James's emphasis on, and attention to, the *processes* of relating makes his version of empiricism radical.

What would it mean to bring this radical empiricist attention to processes of relating into ML and its now-dominant configuration of computational experience? It would mean attending to how technical entities and operations pass, conjoin, traverse, *and difference* via statistical and networked ensembles. The connections concatenated in (any postdigital) computation between entities such as numerical values are discrete; in a neural network, specifically, at the level of its various layers, values are drawn from data inputs and operated on, or values are functionally generated to produce synthetic outputs. Here we have a calculative relation across weights and biases—the parameters for a layer—and data inputs (transduced to numerical values) or other values generated by a random algorithm designed to feed the model some noise. We might say, then, that the calculations performed in a neural network are, at some level, discrete, maintaining the barest relations of "withness" or proximity between the layer and inputs or synthesized values.

These processes occur across one layer with potentially millions of parameters and inputs to produce a new set of values that pass the threshold of that layer to become features, passing to being calculated again until, eventually after many passes, they become the neural network's outputs. Yet such sets of values and constraints calculated together and against one another also easily accord with Lapoujade's reading of the openness of James's experience as a "set of anything that is in relation with something else."

If these computational processes meet the bare criteria of relationality, then they also generate or count in radical empiricist experience. They do not require the appearance of human subjects or objects with their sensory perception or intelligent perception for their operative, calculative experience to "count." This is not the same as saying that ML/AI is autonomous and can run without human control or action. The claims for autonomous AI—whether intellectual, creative, or even functional—are usually premised on the prior existence of the human subject: Can AI be as creative as humans? Is AI more intelligent than humans? Will AI be more efficient than humans by 2050? When using a Jamesian process-based account of ML computation as radically empirical, I seek to approach its operations, sensibility, and relationality differently, suggesting that their relationality counts in contemporary experience. But it also counts for us because its effects register. We only have to think back to Loab and sense that she is such a registration. The proposition for ML set out in this book is for a nonsensuous, liminal, imperceptible, *and* registerable-in-its-effects mode of radically empirical computational experience. This places ML computation/AI and

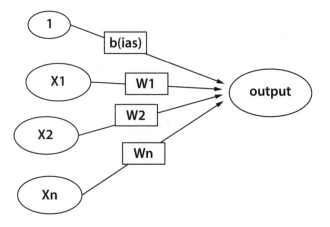

I.5 Schematization of a node in a neural network. The inputs are represented by X1–Xn. A weight, or W1–Wn, is added to these; 1 is the bias. Since all these diagrammatic elements are calculations and sums, this should be understood as a topological diagram of relational values in which the node is a value produced through all these relations.

Pop-Up Definition: Weights and Biases
A node in a neural network—also called a neuron—is a calculative outcome in which a data input (or collection of data points transduced to numerical values) is multiplied by a weight value and a bias value, or parameter. The overall calculation of weight and bias is then further subjected to another "activation" function. The activation function calculates which inputs' weights and biases exceed the threshold of activation (>1), and these are then passed onto the next layer of nodes.
Weights and biases are core calculations in a neural network, yet their activations may be difficult to detect for inputs that are highly variable (high-dimensional data such as images, for example), or in multilayered and massively connected networks. Weights and biases are usually initialized with arbitrary values, and it is the changes to these, and the modulating effects through the network, that become a measure of the model's "learning."

humans together through a relation of *nonrelation* when it comes to knowing and feeling one or the other's modes of individuating experience: we don't know what is happening in the black box of a neural network model; AI doesn't feel creative when Midjourney, a generative AI image platform, makes a new image when prompted with a text input. But it also means that the nonrelation of humans and AI might play out differently when it comes to tallying up what *happens* in computational experience. Relations, whether computationally generated or produced in human-computer conjunctions, are constantly registering in and as experience, and "we" all, together and differently, *experience the experience of* ML *computation*. The pragmatic beauty of James and of process thinking the processes of AI is that we move to interest not in "what is different" about computation (or the same as the human) but in how computation differences or makes both human and machinic experience heterogeneous. Inversely, attending to a radical empirical understanding of ML also allows us to register where computational experience fails at differentiating tendencies as it is harnessed and captured through social, cultural, and political arrangements of predictive computing.

As I have already suggested, ML computation is populated not just by quantitative but also by qualitative processes such as vectorization. We can already see that AI models now performing image recognition and image synthesis, and large-scale language translation and semantic generation are using many of the vectorial relations that James describes as "towards," "against," "because," and "through" via the very logic of their operations. I discuss some of these vectors with respect to image recognition and generation in chapter 1. So, while dealing with discrete quanta, ML processes are simultaneously concatenating operational pathways, and it is these concatenating vectors that *become* the model registering in experience as a dynamic and generative entity. Further concatenations and disjunctions occur via additional processes such as optimizing the model for its specific tasks, as well as human intervention and feedback as part of the model's development, which may be as simple a decision as selecting how many epochs or passes over data a model will run in its training. All these ML and human activities and actions constitute AI as a human-machine ensemble, laying down its relational order of connectivity. In his own time, James offered us a sense of experience being made through conjunctions traversing technical, infrastructural, institutional, and human components: "We ourselves are constantly adding to the connections of things, organizing labor unions, establishing postal, consular, mercantile, railroad, telegraph, colonial, and other systems that bind us and things together in ever wider reticulations.... From the point of view of these partial systems, the world hangs together from next to next in a variety of ways" (1916, 130–31). By paying attention to just such concatenations with respect to ML, as these are made by the operations of models and their *conjunctions* with humans, we can begin to register computation as a singular mode of operativity qualifying our relations of, in, and with culture(s) and technics. James's "experience," I will argue, is well suited to thinking a form of computation, such as ML, whose operations—despite their epistemic and social orientation toward predictive outputs—unfold via continuity and modulation or differencing. And, importantly, processes of differencing also offer contingencies and surprises.

ML's Actual Technics as Computational Contingency

A focus on computation's quantitative and axiomatic potential for novelty alone, which has been pursued in a range of approaches emphasizing computation's contingency, falls short, I think, of the *actual technics* of ML. By "actual technics," I mean two things: first, ML's ensemble of technical components and operations, which can be specific to a domain or operation such as image recognition or natural language production and carry a specific technical lineage through which it gathers its components together; and second, how these ensembles actualize as and through their sociotechnical milieu. As I will explain shortly, this milieu should not be understood as exterior to ML's technicity, where it is "it" that becomes responsible for situating technical elements according to broader epistemic, political, or cultural formations. In Fazi's account of the discrete, quantitative, and axiomatic nature of computation, for example, algorithms do participate in the broader world of social, political, and even material phenomena, but only in a secondary manner in terms of their application and implementation in the world. For her, algorithms are "the a priori intelligible" of computing (2018a, 106). By this, Fazi means that they are mathematical ideas that preexist their embedding into a program or code, and only a particular code or program a posteriori operationalizes them. At the onto-epistemological level, Fazi argues that a deterministic organization of computing operates when aesthetics and logic combine in "computational idealism," in which computational axiomatics are conceived and implemented as the horizon for determining both the ideal/transcendental and the empirical (92). For her, this is how algorithms become predictive or are embedded as forms of governance. But this implies that it is only when computation joins forces with another agenda or when it functions away from its own axioms that it becomes determinate.

But Fazi's conception of the algorithm as a priori axiom cannot account for the kind of computation that constitutes ML and, increasingly, is how computational experience is being made. Machine learning is not primarily a mode of *digital* computation but a statistical one. As I will also argue in chapter 2, algorithms that are part and parcel of ML's operativity germinate from a nineteenth- and twentieth-century *statistical* mathesis in which mathematical ideas are induced from the sociopolitical materialities of race and class relations of Anglo-American nationhood. The distinction between a priori intelligible and a posteriori implementation does not hold for the logic of statistically wrought algorithms. And since ML converges statistics and computation, we will need to look out for its entangling of temporally conditioning oppositions such as "prior" and "post" with respect to data, algorithms, and its entire operativity. Machine learning occurs at the nexus of statistical methods and techniques and computation and is, indeed, a reconfiguration of both. Even if (as we will see at many moments throughout the book) we cannot easily buy into the simple characterization of ML as the emergence of an algorithm from its learning on/of data, nonetheless we must account for a different ensemble in which an inductive technics is at work. Induction, inference, and probability bring different operations, logics, and implications than discrete, a priori axiomatics. As Mackenzie puts it, "Statistics has ... gradually probablized machine learners" (2017, 104). But, as he also notes, ML reconfigures the classical statistical methods of sampling known sets of phenomena such as populations. Instead, ML begins with the premise of operating on all "known" data. Hence ML models are typically thought of as architectures for big data—a dataset of all that can be known. Of course, as Kate Crawford and danah boyd (2012) have already pointed out, claims to the comprehensiveness of big data are limited, since datasets are always in some way historically situated and undergo many processes of organizing and arranging that necessarily filter out data points. Nonetheless, as Mackenzie suggests, the difference between strict statistical samples and ML datasets lies with the latter containing all data "known" for the task at hand. The model itself—in contradistinction to the classical statistician who performed the process of sampling—then becomes the "knower" of the data. This automation of knowing occurs by parsing the data via continuous operations to detect and eventually attempt to eliminate error from its outputs. Mackenzie's point is that ML transposes a probabilistic logic from statistics to the model rather than simply using statistics' *methods*. The consequence of this is the automation of probability. Machine learning delegates the operations of sampling, analyzing, and observing, which account for error and prediction in data, to devices and to the operations of models. For Mackenzie, this also means that the potential for uncertainty—a key and immanent quality to the unfolding of practices of classical probabilistic statistics—is now ceded to recursive elimination performed via predictive operations: "The direct swapping between uncertainty in measurement and variation in real attributes that statistics achieved now finds itself rerouted and intensified as machine learners measure the errors, the biases and variance of devices" (2017, 106). The actual technics of ML is, then, inductive yet also unfolds through a milieu of automated prediction, which it simultaneously enfolds. This fundamentally alters its operative mode of computation from digitality and, I will be suggesting throughout this book, engenders different modes of computational experience.

As we will see, especially in chapter 2 when I look at the ways in which a statistical logic of racialization enters ML, this means that contemporary computational experience can never be easily debiased. The very operativity of this kind of AI runs on a singular trajectory in which techniques or functions of statistical *discrimination* have become immanent to its functioning. Statistical discrimination—through which many baseline algorithms of ML operate—constitutes the actual technics of ML as an already "biased" technical ensemble before any specific data inputs run through a model. Race, class, gender as operations of statistical discrimination become entwined in the core *automated* functioning of ML models. Yet even as ML has automated the project of statistics, it nonetheless remains open to the indeterminacies of probabilistic (statistical) techniques. This occurs regularly in AI models through phenomena such as category mismatches where, for example, images are matched to labels that do not indexically describe them, or when AIs perform in ways that "err" from, yet nevertheless conform to, their task specifications. In chapter 1, I look at ML operations in some detail with respect to image recognition and misrecognition and the ways in which misrecognition recurs across computer vision. My overall proposition—pursued via a close look at a range of computer vision AIs—is that the operativity of ML is not as closed and predictable as is often claimed. Instead, ML is a mode of computation in which indeterminacies are lodged in the operativity of its actual technics. The question will be: How can computational experience remain open to these?

On the one hand, then, we are faced with this delegation of error management—what we might also call the regulation of chance or indeterminacy—to the sociotechnics of predictive computation. On the other hand, its recursive operations, its actualization as part of an ensemble of conjoined algorithms, trained datasets and their legacies, and the contributions of human cognitive and affective interventions in making AI models operative all make ML less predictable. I want to propose that a deepaesthetics of ML must consider both the predictive reshaping of life through automation *and* the potential for new openings onto contingency and indeterminacy. This book engages, then, in a continuous double-pronged approach to teasing out and encountering a deepaesthetics of ML computation: one that recognizes the ways in which relations are reconfigured and often restricted by predictive trajectories; and one in which AI models, data scientists, cultural producers, artists, and theorists are alive to its odd sensibilities and indeterminacies. This requires a thinking of ML intensively and extensively as relational field; we cannot stop at the axiomatic or numerical registers of quanta prehending each other according to algorithmic procedures. We must, however, stay close to the technical specifications of the statistical computation that organizes ML. For it is at the granular level of computation's operativity that we can locate both the production of social propensities, problematic genealogies, seemingly predetermined trajectories and the potential for novel (aesthetic) events and experience.

ML's Machinic Universe and ML as *Agencement*

We should keep in mind that ML's terms, images, and diagrams do not belong to the technical infrastructure of AI alone; they consistently gesture to social, cultural, political, philosophical, and aesthetic ideas and processes. We have already seen this in the case of the layered architecture of deep learning models, which, at the same time, proffers images that tell us about the topological spaces conjured and inhabited by AI. What is particularly telling about the layer diagrams used to explain deep learning neural networks is that they are not images "of" technical components or technical (infra)structures, since the layers are neither physical nor even systemically representative of something technical in the way that a circuit diagram, for example, might be. Indeed, *there are no geometric layers as such* in deep learning networks, but rather only cascading series of numerical values, summations, and operations. But neural network diagrams and the image of the layer as a component of the computational architecture of deep learning computation must not be explained away as an image used to merely "communicate" computation to a nonexpert. The concept of the layer has been operative in ML research from at least the 1970s onward (see Ivakhnenko 1971). In this earlier period of ML research, three terms were used to describe the architecture of neural networks that learn: *hypersurface*, *layers*, and *thresholds*. The hypersurface functioned as a kind of projection of the overall connected topology of the network; the concept of the layer was used loosely to point to the net set of results of known transformations of "groups" of numerical values; and the threshold marked the transformation taking place. Interestingly, layers in contemporary deep learning research merge aspects of all these concepts (e.g., Goodfellow, Bengio, and Courville 2016, 164): they are the convergence point for the model's functions (the earlier layer), the ways these functions interact (the earlier threshold), and the vectorial chain conjoining one function to another (the earlier hypersurface). This later convergence demonstrates the unacknowledged epistemic work now performed by the concept of layer and of the layer diagrams, which often accompany research on neural networks (see fig. I.2).

As both concept and diagram, the layer is thus part of the "machinic universe" of ML, to draw on Félix Guattari's conceptual apparatus for thinking through technologies (1995, 36). For Guattari, any technical machine—a neural network, for instance—is a conjunction made possible through a composition of many elements: diagrams setting the range or vectorial possibilities and constraints of the technical object's feasibility; materialities that enable its production; an industrial sector producing it; a political economy, which finances it; and a collective imaginary that is the ethical and aesthetic condition for it being actualized (48). Taken together, these spheres form a technical ensemble, and it is the dynamic of their relations that brings a particular technical machine into being.[8] While there may be no geometrically shaped physical surfaces, no "places" in a neural network where layers can be located, and no volumetric depth to deep learning AI, layers are nonetheless diagrammatic and aesthetic components of the technical ensemble that is machine learning. They assign its computation a topological architecture and operativity; they also set the limits for imagining what a deep learning AI might be capable of doing.

Yet it is not sufficient to lay out the fields—conceptual, diagrammatic, financial, aesthetic, and so forth—that constitute ML's technical ensemble. Gestures of mapping will not give us a sense of its actual technics, since simply pointing to this array results in a static and almost structural setting in place of ML, or any technical machine, for that matter. Instead, we need a sense of how these spheres come together *in relation*. *Agencement* is the term that Gilles Deleuze and Guattari coined to account for how heterogeneous social, technical, economic, aesthetic, political, biological, inorganic (and more) elements conjoin and multiply in ways that are productive of new relations and events: "An assemblage [*agencement*] is precisely this increase in the dimensions of a multiplicity that necessarily changes in nature as it expands its connections" (Deleuze and Guattari 2005, 8). Like other sociotechnical ensembles, the *agencement* of ML functions by increasing the *multiplicity* of its relations through conjoining with other machines whose dimensions may not be technical at all. This occurs at many different and disjunctive scales: from that of a function such as PCA, a commonly used dimensionality reduction algorithm, to that of the corporate imaginary of AI and its claim for predictive futures. Hence, when I pay close attention to the genealogy and operativity of a technical aspect of ML, as I do many times in this book, I do so to get at the ways in which ML opens onto, conjoins, and enfolds the social, epistemic, political, and aesthetic fields (and more) into it as part of its technicity, or as I have termed this, its "actual technics."

In analyzing, following, and landing on aspects of ML, I contend that computational experience can best be approached via this conception of *agencement*, since it gives a sense of how a technology is always operating within a technical ensemble that is *in process and relation*. The relations conditioning, and the new relations generated by, the *agencement* of ML are experienced not simply by humans but also at and by more-than-human machinic registers. Throughout the book, I retain *agencement* in its original French to set my approach apart from the concept, methods, and framework of what is now called "assemblage theory" (DeLanda 2019). Although there seem to be overlaps with that theoretical project—its emphasis on process, dynamic interrelations, and heterogeneity (Venn 2006, 107)—the assemblage, as the object of assemblage theory, drifts back toward an idea of an emergent system and sometimes promotes methodological scalability. Manuel DeLanda, for example, seeks to provide an entire ontology of social processes by developing a theorization from bottom-up emergent "wholes" such as "persons," understood as assemblages of albeit heterogeneous actions, through to economic or political emergent systems such as finance. This conception of assemblage theory, he states, offers, "an approach in which every social entity is shown to emerge from the interactions among entities operating at a smaller scale" (DeLanda 2006, 90). DeLanda's emphasis is not on processes but rather on entities and how, at each level or scale, one gives rise to or conditions another. This implicitly allows each emergent entity to be brought into relations of equivalence up and down the scale with one another. Perhaps this solves the issue of how to deal with *relating* things that are heterogeneous, but it does so at the expense of the heterogeneities! Instead, ML, as the *agencement* now dominating computational experience, requires an approach that *goes to the processes* activating and making it elastic, the processes that also result in it producing itself contingently. It is these processes of relation that are activators of heterogeneities.

A Simondonian Technics of ML

The concepts of both the technical ensemble and *agencement* in Deleuze and Guattari are indebted to the philosophy of technology elaborated throughout the work of Gilbert Simondon. This book and its thinking through of the actual technics of ML is likewise indebted to many of Simondon's technical concepts. Simondon himself was deeply skeptical about automation, but his skepticism arose not from computation's technicity but from its capture by a sociopolitical reduction of autonomous systems to human behavior and vice versa (2017a, 17). He argued for a restoration of technicity to computation that would acknowledge the "openness" of computational programming: "A purely automatic machine completely closed in on itself in a predetermined way of operating would only be capable of yielding perfunctory results. The machine endowed with a high degree of technicity is an open machine" (17). The sociotechnical program for automated predictability—the publicly declared and dominant agenda for ML in contemporary capitalism—presents just such a closed machine, often yielding cursory results. But this is not all the *agencement* of ML might be or become, and indeed many instances in the actual technics of ML suggest its potential for openness in this Simondonian sense. I turn now to an example drawn from contemporary "automated" music production to see how approaching AI both critically and with a degree of Simondonian openness gives us glimpses of the two poles of deep-aesthetics—predictive and indeterminate—operating in contemporary AI.

The claim that a "deep" *aesthetics* increasingly stakes for automating and autonomizing creativity has gripped the computer graphics and music industries in ways that suggest the *agencement* of ML is always already social, aesthetic, and economic, and more as it actualizes. Popular digitally produced music, for example, has been moving in the direction of deploying algorithmic correction of pitch and vocal timbre through software such as Auto-Tune since the 1990s. While we cannot tell a simple narrative about the uses to which automated digital pitch correction has been put—there are many artful, experimental, and minor configurations of Auto-Tune—the

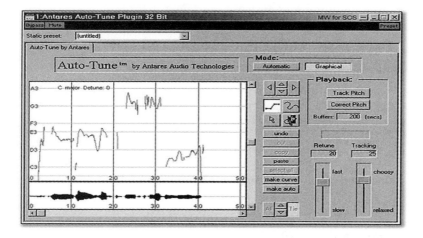

I.6 Early interface for Auto-Tune, ca. 2000, which shows a recorded vocal input signal that could be retuned as its pitch is being tracked. The same capacities are also possible for live performance. Auto-Tune is an example of negative feedback being applied in music production.

Pop-Up Definition: Negative Feedback or Cybernetic Recurrent Causality
Negative feedback occurs when the outputs of a system, process, or operation reenter the system to affect its inputs in such a way as to stabilize further fluctuation in outputs. An example often given is a heating or cooling system controlled by a thermostat. The heated space is kept at a stable temperature by the system if ongoing cooler conditions (cold air entering the space or the temperature dropping owing to moisture, proximate conditions, and so on), detected as inputs to the system, are modified by the thermostat. This would result in the ongoing "output" or room temperature being kept at a constant warm temperature despite new inputs varying.
Negative feedback in cybernetics tends toward stabilization of a system by closing the difference between new information or variability and a continuous output of signal modulated by the system's internal operations.

assertions housed in its original patent nonetheless underscore the homogeneous tendencies that it has cultivated since its release in 1997: "Voices or instruments are out of tune when their pitch is not sufficiently close to standard pitches expected by the listener, given the harmonic fabric and genre of the ensemble. When voices or instruments are out of tune, the emotional qualities of the performance are lost" (Hildebrand 1999, 1). Here we move in seamless fashion from performer to pitch to listener to emotion, all to be navigated via an algorithm that performs an operation of *standardization*. The Auto-Tune algorithm works by automatically detecting an actual human performatively generating pitch; sending inputs to a corrector (a set of MIDI standardized pitches), which automatically correlates these to match the standardized pitches; and then re-outputting the performed pitch as corrected or retuned.

This input–correlation–output cycle is one of the simplest schemas for automating human-machine relations and falls within the shadow cast of what Simondon called cybernetic "recurrent causality," or what we more frequently call "negative feedback."[9] Here, (first-order) cybernetic design creates a link between "the chain of causality conveying the action and the chain of causality conveying the information" (Simondon 2020, 427) by literally capturing the latter (information) in a circuit for (re)producing the former (signal) as the key to the operativity of its system. For Simondon, this means that whatever is potentially novel about the information—whatever is contingent, in other words—is discarded to ensure the ongoing hegemony of seamless signal. In the operation of the Auto-Tune algorithm, the only information that comes to count is *the difference between the pitch produced by a subject and the pitch that needs to be corrected/produced by the software*. Consequently, the subject/performer/musician learns from that narrowed range to adapt their performance and behavior, linking their ongoing action/performance to receiving that correction alone. Accordingly, the kind of sound produced tends toward increasingly correlating the sung voice to the software's processes of standardization. Over time, a homogenization of vocal sound occurs unless other variants of information are introduced and explored. The issue here is not the dominance of "the algorithmic" per se but rather the privileging of operations of causal recurrence, which divest the human-machine relation of a broader milieu of variable information. The dominance of cybernetic recurrent causality or the paucity of generating multiple kinds of information for software and hardware systems has primed a good deal of music production and performance for further regularization via ML.

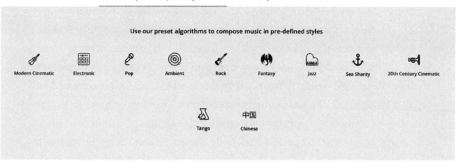

I.7 From AIVA's (a music composition tool using RNNs) website. This allows choosing preset styles that contour the music as it is made to fit into genres and even culturally specific music styles on which the model has trained.

Pop-Up Definition: Recurrent Neural Networks (RNNs)
A **recurrent neural network (RNN)** is a type of neural network architecture used to learn from sequential data. Examples include letters in a word, words in a sentence or sequence of text, and musical sequences. A key characteristic of their operation is that data are inputted as vectors or ordered strings of numbers. These vectors can be stored in layers and used to sum, multiply, or subtract with another vector that is fed back into that layer. This allows, for example, predictive text to use other sequences of words (other vector states that have been stored) that supply contextual information to make correct "guesses" or predictions.
Recurrent neural networks are core architectures in the development of natural language processing. While now superseded by other architectures, they have provided the programmatic conditions of possibility for large-scale language text-to-image models to emerge. These use a network architecture called a transformer building on early research in which two RNNs were used together (encoders and decoders).

Enter Amper, IBM's Watson Beat, and Google's Magenta, deep learning models that are being used widely in computational music production. Although these are not stand-alone models—Magenta, for example, leverages TensorFlow, now a subplatform within Google for developing many kinds of ML algorithms—they all use recurrent neural networks (RNNs) or variants of them. Recurrent neural networks are models often used for processing data with sequences of values occurring over time. They store values as vector states—a sequence of data inputs transduced to strings of numbers—which subsequent states can connect back to iterate on. At a much larger scale and using many more recursions, they apply principles of feedback loops. Their neural training for music production usually occurs by feeding in many sequences of melody, chord, or rhythm progression data until the model learns the style of the music to be generated. Whereas in Auto-Tune the recurrent causality operates across software and human voice, in RNNs the recurrence works on preexisting digital samples of musical style and genre, which may be audio recordings or MIDI files. Importantly, this already suggests a transduction of the performative information, entailing the model using samples that have already been compressed or even pitch corrected by digital preprocessing algorithms. The RNNs can synthetically (re)produce genres and even moods as presets for musicians learning from these samples to generate new musical refrains, percussion sequences, and so forth. Crucially, then, at different times and scales, the models are learning data *already schematized* by processes of recurrent causality, embedded in the samples and widely circulating via types of signal processing—including Auto-Tune but also digital filters and effects, for example—and formatting (the ubiquitous MP3 file).

What, then, are the role and place of a human musician or performer working with such an AI if a certain schematization of human-machine activity has already been subsumed into the model? In what ways might human vocal performance introduce variable information or be subsumed into a schema of recurrent causality? We can look, first, to the example of the pop singer Taryn Southern, who released the album *I AM AI* in 2018 using both Amper and Watson Beat. A former star of *American Idol*, Southern had previously worked with music producers to create "structure" for her pop songs (Deahl 2018). Using AI music platforms, she chose genre and mood presets to generate a compositional base for her songs and then iterated melody and lyrics over the top of them. This is not so different from compositional strategies for many contemporary pop and electronic musicians, who use a range of digital audio workstations such as Apple's

Logic, which already integrates various AI operations into its software. But what is revealing is Southern's attribution to the AI of both compositional and cognitive agency as it learns preprocessed *samples*. At the same time, she gives her own musicality a programmatic character by describing her efforts as iteration or *recurrence*: "She knew 'very, very little about music theory.... I'd find a beautiful chord on the piano... and I'd write an entire song around that, but then I couldn't get to the next few chords because I just didn't know how to play what I was hearing in my head. Now I'm able to iterate with the music'" (Southern, quoted in Deahl 2018). Here we might say that the human slots into a diagram of doubled recurrent causality in working with the AI's predictive operativity. Since the model is already working in a recurrent mode deploying presets, the musician/singer portrays their melody as a pattern of recursion on top of the underlying recurrence being automatically generated. Here recurrence *recurs*, intensively within the model's own processes and extensively by drawing the human into its diagram. In this respect, the human-machine relation is configured so that all elements, human and technical, are seen to be functioning according to a reduction of information and variability to seamless signal output: AI presets and human iteration match each other. Any openness to something outside the programmatic, or what Simondon also calls a machine's "margin of indeterminacy" (2017, 17), is foreclosed. A fully automated mode of computation has no margin of indeterminacy, since theoretically it never varies from what it is predicted to be or to deliver. Variability and hence novelty or new information—the musician/singer, for example, improvising with the AI, and the AI varying and modulating because of that improvisation—could only occur in a technical ensemble that is also open to the unforeseen (39).

If we take an altogether different approach to working with an AI as an "ensemble member" within musician Holly Herndon's practice, we begin to see how she engages a sensitivity to a form of recurrence that is variable and contingent.[10] Likewise deploying RNNs, Herndon developed a combination of spoken and sung extended vocal audio samples from her own and other female singers specifically produced to train her AI, Spawn. An artful technique is here doubly articulated with respect to cybernetic recurrent causality: both refusing to attenuate her own voice to a predetermined set of canonical musical expectations such as standard pitches, and rejecting attenuation of the voice of the AI to pregiven data—even her own prior recordings— that could be mined for a style. Consequently, early experiments with Spawn generated pitch and rhythm continuity with the extended vocal technique tendencies of Herndon's music making but also developed variable machine-generated vocalizations. These sounded like a supra-beatboxing mode of vocalizing in which rhythm—rather than oscillating between the regular and syncopated—became irregular and contingent. Herndon (2021) notes that she didn't train the model to beatbox per se, that is, to explicitly generate synthetic vocal percussive sounds that in their live human form aim to imitate drum machines, samplers, and the rhythmic components of hiphop. Instead, something analogous to beatboxing emerged from Spawn's synthesizing outputs by learning patterns of feature distribution across sung and spoken audio sequences made for the training dataset. Spawn's vocalizations have no consistent, stratified, or predictable beat. Instead, an echo of the style we name "beatboxing" is conjured as a musical gesture by the AI. As Herndon suggests, strict digital sampling such as simply taking a riff, rhythm, or vocal phrase and replaying it within another piece of music repeats past musical gestures, whereas "spawning builds variations from past expressions" (2021, 45:45). Machine learning here is artfully engaged by enabling differential recurrence via spawning rather than reproducing known expression of styles and genres. Alfred North Whitehead calls this the production of "intense experience," in which a graded set of contrasts emerge in relation to identity (ground, past expressions), ushering in the emergence of novel "aesthetic fact" (1978, 279). In Herndon's co-composition with Spawn, collective registers of musical phyla—extended technique, digital audio processing, and beatboxing—variably condition the performance of the individual human musicians and of the musicality of the AI.

The overall outcome of the Herndon-Spawn relationality is a recurrent vocalization at odds with itself—slightly out of phase rather than predictable. I will look to the ways in which artistic engagement such as Herndon's pulls at, twists, and cajoles the schemas that sociotechnically organize the overt deepaesthetics of ML. Careful and artful techniques can cut into, conjoin with, or jam open a margin of indeterminacy for AI. Artists, in this capacity, are not those who represent deep learning as either humanlike or transhuman, nor are they interested in drawing out its creative agency. Instead, they are the conjurers and crafters of artful techniques: artists, cultural producers, critical thinkers, and data scientists alike. These techniques and this "artfulness," as Erin Manning (2016, 46) calls art practices' capacity to generate new opportunities for relation and for living, produce opportunities for becoming sensitive to a "what else" for AI. Here Simondon's conception of what a thoughtfully deployed imagining of technics might do is as useful for ML as it was for the thermodynamics and cybernetics dominating the twentieth century: "We can consider the technical imagination as being defined by a particular sensitivity to the technicity of elements; it is this sensitivity to technicity, that enables the discovery of possible assemblages" (2017, 74).

How Else for ML?

Throughout this book's four chapters, artful techniques surface that bring the ML operations and deep learning models being rolled out in contemporary social, medial, and political contexts into new relations with their technicity. What these techniques hold in common is both a sensitivity to the actual technics of ML and a desire to ask: How else might AI become even amid its trajectory toward prediction? This requires staying close to "the technicity of elements" but also seeking out what Guattari calls the "allopoietic" dimension of all systems and processes (1995, 47). This is the collective dimension of alterity with which any entity or system is always already in relation, and which enables novel generative capacities. To consider organic life momentarily, genes, seemingly units that underpin the self-production of an organic system, can only reproduce life by carrying the potential mutations of their entire genetic phylum. The gene is not an isolated unit, then, but immanently retains a past of actual changes and hence the potential for future changes. In this past-future/present-past/present-future topology lies the gene's dimension of and for expressive alterity. Technologies too, while inorganic, immanently carry their pasts—the phyla of their realized and *unrealized* sociotechnical mutations. They too open onto other futures. These potential lines for unfolding provide conditions for different ways they might unfold than their current realizations.

In the first half of the book, I alight on two prominent areas of computational experience: automated image generation via deep learning in chapter 1, and the racializing potential of statistical techniques in chapter 2. My project in these chapters is one that engages closely with the actual technics of ML, remaining sensitive to the potential for an alternate technical imagination and the conditions for another becoming for AI. In chapter 1, I discover the ways in which experimentation with the category mistakes of computational vision takes deep learning models toward potential indeterminacy and an alternate deepaesthetics. In chapter 2, however, the actual technics of statistical racism and its increasing manifestation in social media, in training data and trained models, and in the algorithmic politics of all of this warrants a retraversal of the machinic phylum of ML. I focus on the two statistical techniques of PCA (which we looked at earlier in the context of dimensionality reduction) and linear discriminant analysis (LDA).

By tracing the immediacy of their relation to the social program of eugenics in the nineteenth and twentieth centuries, we can see that these techniques *operationalize* race in a manner that is key to the sociotechnical ensemble that becomes ML. It is not simply, then, that datasets are racially biased (or biased against other "recognizable" social groupings). Instead, statistics, in its deployment of discrimination techniques in a large-scale automated mode—that is, platform-enabled ML—shapes and contours data so that they are distributed toward whiteness and away from Black, Brown, Yellow, or other kinds of "color" experience(s) of life, race, and bodies. Here, again, we could say that the *agencement* of ML closes computational experience to any variation that is not already predetermined along a spectrum distributed according to whiteness as its normative curve. What would it take to *artfully* prize open this deeply racist aesthetics to other kinds of experience? Throughout chapter 2, I visit the work of Stephanie Dinkins, whose proposal for an "Afro-now" AI lays out just what it might take to really generate different color spectra for computation. In Dinkins's practice, we see a refashioning of ML as a different kind of assemblage, whose conjunctions hold through relations of experiencing and experiences of making Black and colored life through familial socialities, together with participatory design engaging Black and colored communities of (technical) practice. Against the racism of statistics' eugenic genealogy and platform AI's deepaesthetics, Dinkins deploys deep learning models as open, contingent, nonscalable, and differentiating ensembles.

The operative rhythm of chapters 1 and 2 moves according to a pulse that first locates the potential of ML's machinic "error" (computer vision's category mistakes) and then genealogically traverses the very operativity of statistical computation as *agencement*. Chapters 3 and 4 invert and convolve the maneuvers of the book's first half, accenting what is often cast—for example, in speech pathology—as a disfluent and neurodiverse "outside" of language as condition for the possibility for the technical ensemble of *natural* language programming. Then, in chapter 4, I return to how contemporary artists engaging ML tease out its potential as *errant agencement*. In chapter 3, I am keen to show how AI's fashioning of "natural" language ontogenetically leans on what is disavowed in the quest to make its artificial agents speak seamlessly. This disavowal involves both an incorporation and a denial of nonlinguistic disfluent aspects of language production. The actual technics of AI language agent development enfold the affective and neurodiverse asignifying conditions of linguistic sense making so that these become the conditions of possibility for an AI-human conversational relation. If race is the unacknowledged *interior* core of statistical operativity and hence propels the functional continuity of ML models, then the *exteriority* of (normalized "fluent") speech—disfluent and neurodiverse communication such as stuttering—is what affectively enables the project of a *natural* AI-generated language. By the end of the book, having traveled back and forth across ML as *agencement*, I return to the generativity of the mistakes made as ML performs functionally or on task. I suggest these are less category mistakes to be fixed by recategorizing, and more a mode of persistently erring that is lodged in ML's operativity. This errant mode is being drawn on by a range of practitioners interested in fashioning a critical sensibility for AI via techniques of artful modeling, crafted datasets, and sensory and insensible experiments with ML operations and processes.

Together, these and related techniques begin to constitute an artful mode of knowing the operations of a technical system, which Simondon (2020) called "allagmatics." Here, analogically tracing and enacting system operations dynamically reanimate the (structural) elements of a technical ensemble, resituating how the system's technicity might be made known, "by defining structures based on the operations that dynamize them, instead of knowing by defining operations based on the structures between which they are carried out" (Simondon 2020, 666). The artful techniques devised by artists who critically engage ML reperform the operativity of AI yet simultaneously diverge from its homogenizing predictive structuration. Instead, these techniques analogically enact *differential* repetition of the contemporary *agencement* of machine learning. Their artfulness lies in the extent to which the artworks retain their margins of indeterminacy between AI's procedural playing out within current sociotechnical constraints and a carefully considered deploying, a making errant of AI that subsequently makes it wander away from being "on task."

Heteropoietic Computation

Category Mistakes and Fails as Generators of Novel Sensibilities

In a research project experimenting with machine learning (ML) models and functions on unlabeled images drawn from scientific research papers and publications, I became fascinated with how frequently standard image recognition algorithms were at a loss to classify visual forms deployed by data sciences such as computer vision.[1] The project involved building a large dataset of scientific images drawn from all preprint articles deposited in the open access repository arXiv from 1991 (its inception) until the end of 2018.[2] ArXiv holds research papers across a range of scientific knowledge domains including physics, mathematics, statistics, computer science, and subfields of biology and economics. Its development across three decades has happened in tandem with a number of key changes in the ways ML has inflected images and image-related tasks: the release of the MNIST handwriting dataset in 1998, which became a benchmark for the visual recognition of handwritten numerals in particular; the creation of ImageNet in 2009; Google Brain's 2012 deep neural network architecture AlexNet, which became famous for recognizing cats in unlabeled images taken from frames of YouTube videos; and the generative synthesis of images using generative adversarial networks (GANs) from 2014 onward. All these significant developments in AI image production and the use of ML as a visual technology are championed and deployed in thousands of research papers deposited in arXiv with accompanying images, often produced through ML techniques. Yet when we used a standard ML classifier, VGG16, which is a convolutional neural network (CNN), attempting to "learn," label, and predict how ML itself might "look back" at images of its own making, the results we returned were seriously out of whack. Where the scientists carefully captioned their images as types of scatter plots, trajectory predictions, heat maps, and segmentations, the classifier predicted these to instead be "oscilloscopes," and "nematodes."[3] Rather than "recognize" entities from their own machinic phylum, these benchmarks of ML-based AI performed spectacular category mistakes often on images generic to their own mode(l)s of production.

When the twentieth-century British philosopher of mind Gilbert Ryle named "category mistakes" as the root cause of the mind-body problem (2009, 11–12), he was at pains to set out how terms should be separated into their correct classes or logical types and not ontologically mismatched. Category mistakes are typically illustrated by sentences such as "The theory of relativity is eating breakfast" or "Colorless green ideas sleep furiously" (Chomsky 1956, 116). While grammatically correct and even statistically plausible (Pereira 2000), a category mistake is characteristically odd and inappropriate rather than false or grammatically incorrect. Although not named as such in the philosophical literature, the duck-rabbit illusion problem, appearing as an illustrated image in *Fliegende Blätter*, a German weekly, in 1892, is a perceptual corollary, insofar as it visually demands that two categories of sentient life—birds and mammals—are able to be visually swapped out for each other. Categories and their mistakes, then, seem to pervade and inflect dominant paradigms of cognition and perception from the nineteenth century onward. Could it be that the category mistakes of ML-based AI, especially pervasive in computer vision, foreground a kind of strangeness at the heart of both the operativity and sensibility of contemporary computational experience?

Kaninchen und Ente.

1.1 "Rabbit and Duck." First known illustration of the optical illusion published in the German weekly *Fliegende Blätter*, 1892.

In *The Concept of Mind*, Ryle points the finger at "foreigners," whom he asserts are guilty of committing category mistakes, burdening them with misunderstanding fundamental categorical differences embedded within a native speaker's linguistic arsenal. Ryle's examples of the performance of categorical errors by foreigners turns out, unsurprisingly, to be due to their unfamiliarity with English class-based institutions such as cricket and the Oxbridge system. Being shown around the grounds and buildings of Oxford, including its offices, teaching spaces and residential colleges, the foreigner asks, "But where is the University?" Again, the foreigner watches a game of cricket and understands scores, player roles, and umpiring yet states: "But I do not see whose role it is to exercise *esprit de corps*?" (Ryle 2009, 17). Ryle attributes cultural naivete to foreigners but argues that they commit *logical* mistakes. The university and team spirit are mistaken as physical instances, whereas their reality resides elsewhere in the *meta*categories of structural organization and sporting performance. For Ryle, the foreigner has logically confused substantial things, places, roles, and tasks with abstract concepts, shared knowledge, and schemas.

In what may be a full circling back to the category problems posed by optical illusions and Ryle's problem of logical confusion, the most sophisticated contemporary AI image models also struggle with generating optical illusions and with picturing metacategorical concepts of, for example, universities as institutions of knowledge production. Designed as paeans to category *recognition,* it seems odd that text-to-image platform such as DALL-E 2, Stable Diffusion, and Midjourney, which should have emerged as the successors of the last two decades of image recognition ML, frequently perform in ways analogous to Ryle's foreigners: swapping categories for each other and taking the physical as representative of the abstract. Figures 1.2 and 1.3 show two different image outputs generated by Stable Diffusion and DALL-E 2, respectively, both publicly accessible, text prompt-based AI image generation tools.

These kinds of tools have become part of the debate about the creative agency of AI owing to their capacity to easily output high-resolution images in styles that range from 3D game-engine computer graphics to impressionist brush strokes. They have been lauded as models that will have "significant, broad societal effects" (Ramesh et al. 2021) and, especially within the computer graphics industry, as a "game-changing technology for art" (Boney 2022). In figure 1.2, the fundamental imaging problem with which the model grapples *is* the optical illusion. In the original illustration of "Rabbit and Duck" (fig. 1.1), visual perception both stabilizes and destabilizes across the relation to the figure as a whole: perceiving a duck, rapidly switching to a rabbit, or adamantly refusing to transpose from one category of sentience to another. The image generated by Stable Diffusion (fig. 1.2), on the other hand, "resolves" the illusion into a morphological confusion of a rabbit shape-shifted into a duck while simultaneously multiplying bits of either's body parts. We don't see two figures in the AI image; rather, we are presented with a nonexistent "both" creature, which strangely gestures toward its own self-generation dispersed across its surrounds. The optical "illusion" cannot be generated by the AI, since the illusion exists *perceptually*; that is to say, it exists in the relation between image and a viewer with eyes (in consort with other senses such as proprioception) and not simply in the image itself.[4]

In figure 1.3, DALL-E 2 is guilty of the same mistake as Ryle's foreigner: a "university" is reduced to an amalgam of built facades, headed by creolized Latin text deriving from the model's distributed sampling of institutional mottoes, with this the only visual gesture to its medieval genealogy remaining. What we know about the training of text-to-image models, however, is that a corpus of matched text-caption and image-caption pairs have been gathered by developers, scraped from social media and other internet sources (such as Wikipedia).[5] "A university," then, must be visually synthesized by DALL-E 2 from a distribution of images of universities originally

1.2 Image generated using Stable Diffusion with the text prompt "rabbit-duck illusion, in the style of a lithograph."

1.3 Image generated using DALL-E 2 with the text prompt "a university."

> ***Pop-Up* Definition: Text-to-Image Models**
>
> A **text-to-image model** is an ensemble of neural networks that input natural language text and output images probabilistically referencing the text input. The main models supporting this ensemble are large language models (LLMs) and generative image models. The possibility for text-to-image conversion comes from the large amount of training data used for the LLMs, scraped from online sites where images are already labeled, such as Flickr and Wikipedia. This allows the models to train on data in which image and text are already matched, providing the basis for multimodel correlation and then image generation that "matches" text descriptions.
>
> **Text-to-image models** gained widespread attention with the scaling up of language models and the generative image capacities demonstrated by DALL-E in 2021, running on the OpenAI platform. The subsequent profusion of generative tools and platforms such as Stable Diffusion, Midjourney, and the ERNIE bot has sparked a debate about the artistic possibilities and ethical pitfalls of AI. Some of these issues include questions of who generated the original unacknowledged and unpaid training images, and whether text-to-image, video, and more AI will replace the need for human graphic design.

scraped from online sources; no doubt this includes many images from actual universities' own websites featuring their facades along with examples of capital works carried out on their grounds. The overall effect in the DALL-E 2 image is to visually render "the university" increasingly indistinguishable from other corporate organizations, at which figure 1.3 hints. Its brightly bricked grid with embedded reflective glass windows would not seem out of place in a business park. Perhaps, then, the AI's image generation performs the reality of a category confusion that already besets the contemporary university within neoliberalism—precisely that a university's image merges with a facade of corporatism. In some ways, the probabilistic operations of the model's image generation, which I will examine in a moment, act as a diagnostic of the social imaginary that is the milieu of today's institutions for knowledge production.

While predictively inaccurate as a descriptive or indexical image in both these instances, the AI performs something interesting nonetheless based on the operativity of its *agencement*. As its name suggests, these AIs are image and text deep learning networks; they function within a technical ensem-

> ***Pop-Up* Definition: Latent Space**
>
> **Latent space** is a mathematical nonphysical distribution of data points into relations of resemblance, achieved by the operations of machine learning models. *Latent* refers to the nonvisible and nonextensive nature of the "space." The closer the resemblance of features being trained for and by the model, the more proximate their data points will be in a latent space. For example, the data points of image inputs that are mostly black and white would be more closely clustered together than images that also include a random array of colors. Latent space is an outcome of both a model learning and extracting patterns of features across data and the processes of that training. In many cases, the redistribution of data points via patterns of resemblance also involves a compression or downsampling of the amount of data points inputted into a model.
>
> **Latent spaces** are said to be hidden because their precise operationality is not necessarily discernible to, or interpretable by, human perception or cognition. Nonetheless, they have become a source of artistic experimentation and are featured in what are known as walks through latent space in the morphing images and animations of, for example, Robbie Barrat, Mario Klingemann, and Anna Ridler.

ble, harnessing the capacities of what are known as diffusion models. These are probabilistic neural networks that train on the latent space of a dataset rather than on raw data inputs such as image's pixels. In chapter 2, I spend some time with the concept of latent space, since it helps shape the ways in which a model skews toward certain features over others. As we will see later in chapter 4, latent space has also become a site of much artful experimentation. For now, the simplest way to think about latent space in relation to the raw data inputs for a model is to conceive it as being a lower-dimensional set of compressed features of data existing in a specific pattern of distribution across hidden layers of a neural net. Rather than all the information or data points from all inputs, latent space retains only the main features from the data inputs. Just as we saw with dimensionality reduction in the introduction, latent space is also a kind of reductive representation of data's features, compressing data points into a new configuration and vectorization, losing irrelevant features and stabilizing the data into patterns of similarity. In a diffusion model, the distribution pattern of these compressed data points throughout the latent space becomes the training data for the model rather than all the raw input data points. What is important to note is that these models are learning the patterns of similarity distribution of data in this latent space, with other dimensions of the data discarded as "imperceptible details," such as a small misshapen edge or a noisy pixel from an image: "Compared to the high-dimensional pixel space, this [latent] space is more suitable for likelihood-based generative models, as they can now (i) focus on the important, semantic bits of the data and (ii) train in a lower dimensional, computationally much more efficient space" (Rombach et al. 2022, 4). Here, the "semantic bits of data" are privileged as those features in an image that are recognizable: shape, edge contrast, foreground, background, and so on. But we should also remember that text-to-image generators are "likelihood-based generative models." Hence, while the text prompts fed to the post-GPT-3 image generation models tend to produce highly descriptive and nominalist pictures of recognizable "things," their imagistic performance operates in more indeterminate ways amid zones of "likelihood." You can feed the same AI the same text prompt at different times, and it will produce different images, for example. Inevitably, the AIs also fail, and there has been much discussion about their inability to accurately render groups of things, text, and graphic writing in images, and their response to prompts with ambiguous syntax.[6] But these fails—and especially the rendering produced by a rabbit-duck illusion prompt instruction—tell us about the ways in which the model becomes inventive in the face of category breakdowns. Instead of presenting two simultaneous categories occurring within the one perceptual event, "rabbit-duck illusion, in the style of a lithograph" produces a third category that is neither rabbit nor duck but more than both. These fails also suggest how AI-generated images might be deployed as a critical diagnostic of the image ecologies—image as brand, image as data resource, image now functioning only in an ecology of other images—that pervade computational experience under the dominion of ML.

Nonetheless, much data science strives to uphold category discreteness. Ryle's attempts to systematically separate classes of entities from each other is matched, even superseded, today by the efforts by which data, algorithmic process, and AIs are worked over to likewise mitigate against category confusion and the threat of computational dysfunction. Just like Ryle's characterization of the "foreigner" as the one who transports ignorance into the operations of categorization, the confusion cast by misrecognition in AI needs to be jettisoned, and accurate vision established instead. Yet a foreigner's querying might afford us traction in understanding the epistemological politics at work in categorization as a form of privileged and highly closed knowledge production. I want to propose that unlike Ryle's project for eliminating the category mistake by naming it a foreigner's error, its occurrence in AI image cultures is indicative of its immanence to ML as technical ensemble. The category mistake is the foreigner, outside, and alien internal to an aesthesia of computation reconfigured by probabilistic statistics. As it manifests persistently across the development of computer vision, it has increasingly bestowed an infelicitous affectivity diffusing throughout ML's predictions, artifacts, and models.

From Category Mistakes to Computational Heterogenesis

Artificial intelligence is made operable by a statistical mathesis, in Michel Foucault's sense of the term *mathesis* (2002, 80–81), as an ordering of attributes and judgments according to equivalences, to determine "truth." This statistical mathesis organizes AI as a mode of knowledge production in which the accuracy of data in relation to an external referent or empirical index is less important than the "accuracy" of a training set on which a model or algorithm learns. Here *accuracy* means the extent to which the dataset adheres to accepted statistical standards: Is it large enough for the task at hand; has it been managed or adjusted for error margins; does it replicate commonly accepted data standards; and so forth? The logic of such a statistical mathesis unfolds as a forceful tendency toward generating equivalences across data inputs and predictions such that these "match" each other. And here is where we can likewise find attempts to optimize a model so that it no longer performs category mistakes. This pull of AI's sociotechnicality in the direction of smoothing over its infelicities inevitably results in a banal deepaesthetics. Critique of various fails and misrecognition instances in creative deployments of computer vision, for example, often centers on models not rendering "accurate" descriptions or staying on task and not yet being on par with the perceptual or linguistic development of a human child. Yet this gets us no closer to what kind of sensibility is generated by and with human-AI encounters, and tends to diminish the potential inventiveness of ML computation. Artificial intelligence instead needs to be poked with concepts such as Guattari's "machinic heterogenesis" (1995, 33), in which its technical ensembles are opened along technological, semiotic, axiomatic, and social axes to processuality. This would allow us, Guattari argued, to "discern various levels of ontological intensity" (34) and to grasp, with respect to particular machines, the singularity of their modes of expression. Today such heterogenesis occurs as quantification and calculation crisscross the *agencement* of platform, infrastructure, computational form, and sociotechnical configurations in asymmetrical, recursive, and nonlinear ways. This provides the potential for computation to become other than a sociotechnical machine for equivalence. The category mistakes and fails of ML signal this very processuality of computation: its potential for heterogeneous opening to different sensibilities.

Nonetheless, and despite scenarios of a transhumanist future, a generalized risk-averse cultural imaginary is at work for AI. Both humans and AIs are frequently reduced to ideas and images of each other. The deeper or more layered the neural network, the more like some aspect of the human mind it is deemed to be. Take Google's DeepDream, for example. Here the network iteratively trained to find features such as the proximity of certain colored pixels to each other, fed them forward through the layers, and then reapplied the learned features back into randomly generated noise images (Szegedy et al. 2015). The features of interest for DeepDream—or rather the ways its network nodes were weighted and biased to favor one colored pixel in proximity to another, for example—were originally learned by training on a subset of the ImageNet dataset. The ImageNet dataset project was initiated in 2009 for the purposes of advancing computer vision research (Deng et al. 2009) but by the following year already had an offshoot in its ImageNet Large Scale Visual Recognition Challenge (ILSVRC).[7] The speedy interpolation of the dataset with its sociotechnical milieu of computer science researchers, students, and corporate tech platform partners (Google, Nvidia, the National Science Foundation, Princeton University, and Stanford University, for example, were all involved in the ILSVRC) competing to create image recognition algorithms is important to note, since it highlights the parallel tendencies at work in ML toward equivalence *and* multiplicity. On the one hand, the ILSVRC standardized a benchmark subset of ImageNet by releasing only one thousand of its image categories as the challenge to be worked on by a model toward accurate classification. This meant that the same categories were distributed and recirculated across image recognition research practices for eight years—a reasonably stable amount of time for a fast-shifting technical area. It is this paring down to a subset of one thousand categories, albeit with some internal differences from year to year, that signals practices of standardization underpinned by an implicit desire for sameness. On the other hand, understanding ML here as *agencement* lets us see the technical elements of AI, such as datasets, as much more than just computational infrastructure but part of a social effort of communities of practice through which they are made into benchmarks. Here, then, multiple machines for producing AI are at work: social, technical, cultural, and epistemological.

For 2014's ILSVRC, Google's DeepDream generative image AI trained on an ImageNet subset, which included things, beings, and objects from English red setter dogs through to dumbbells.[8] The trained model, known as Inception, which I briefly touched on in the introductory chapter, was then given new input images of randomized noise in which no humanly discernible representations were present. Through multiple passes of these noisy images across its network layers, DeepDream effectively overdetermined the noise as fractalized fusions of previously learned features from its one-thousand-category subset of ImageNet, which for that year contained numerous dog breeds. Google researchers somewhat playfully referred to the synthesized image outputs as neural net "dreams" (Mordvintsev, Olah, and Tyka 2015), and the resulting images were at once likened to the stuff of hallucinations and declared "artistic" (Levy 2015). Machine learning seemed to have suddenly acquired those all too elusive human qualities of both creative capacity and unconscious life. In the viral dissemination of DeepDream images that followed—via DIY online image generators as well as notable "aesthetic" events such as the 2016 art auction of DeepDream image outputs at San Francisco's Gray Area Foundation for the Arts—deep learning was deemed to be acquiring its own "imagination." While DeepDream's imaging may have been temporarily dubbed computer dreaming, it was designed to generate pictures that eventually iterated from noisy inputs to represent objects in the world like bananas and buildings: "Say you want to know what sort of image would result in 'Banana.' Start with an image full of random noise, then gradually tweak the image towards what the neural net considers a banana.... By itself, that doesn't work very well, but it does if we impose a prior constraint that the image should have similar statistics to natural images" (Mordvintsev, Olah, and Tyka 2015). The machine "psyche," then, emerges from the operations of the AI's own computational processes of learning, modeled according to a representationalist paradigm of (visual) perception widely deployed throughout deep learning (e.g., Goodfellow, Bengio, and Courville 2016, 353–56; Hinton 2006), statistical operations (here acting as constraints, which we will later come to consider as parameters of the model), and the social milieu of the ILSVRC, which enfolds into its technical ensemble.

However, the resulting "psyche" and pareidolic images, which captured public attention, seemed to indicate a creative, imaginative AI with a mind of its own. DeepDream's dreaming appeared to emerge autopoietically, as if something with both representational realist *and* surrealist tendencies were developing from the bottom up. Yet rather than a sense of machinic alterity, and a potentially heterogeneous unfolding, DeepDream is instead granted (autopoietic) autonomy. And this brand of autonomy takes normative human experience—in this case, visual perception modeled on representationalism with an epiphenomenal capacity to "hallucinate"—as the benchmark for computer vision. This ensures that the complexity of the human-AI relationality that makes up the sociotechnical ensemble of the DeepDream model does not garner attention. What is lost or missed in all of this is the possibility of encountering and moving relationally together or apart, the possibility of other kinds of contemporary technical experience that might allow for the heterogenesis of both human and machine learning. One strategy for refusing such reductions is to insist on AI as neither seamless nor error prone but rather *at odds* with itself. Returning, then, to the idea of the category mistake as something equally intrinsic *and* inimical to the operativity of ML is to insist on the ways in which AI might be conceived and practiced as a differencing machine. While beginning with the grip that equivalence exercises on machine learning, we might also seek out where one already sees AI's statistical mathesis diverging from itself, the ways it transduces quanta as qualitative processes that register via odd, inappropriate, and even inapt functioning.

Calling on Guattari's machinic heterogenesis and Simondon's margin of indeterminacy, I think we can develop a relational thinking for and of ML that is generative of different kinds of computational experience. This shifts emphasis to the novel and strange potentialities of models as they operate diagrammatically and topologically, and as they inaugurate novel possibilities for machine-human encounter. Allopoietic rather than self-replicating, AI is dynamic and can be productive of something "more than" the human, to which it is constantly held as benchmark. It positively incorporates or conjoins with elements at, from, and of its indeterminant margins—or even at its core, such as the category mistake—and transduces those elements in divergent directions that are nonpredictive. Opening ourselves to the wider milieu of its (socio)technical *associations* (and not simply to its other technical element*s*) may just make it possible to envisage a different deepaesthetics for ML-based AI.

Constraining Computer Vision

At the same time, I want to be circumspect about the very real constraints of the predictive and autopoietic vectors that close down what could be open and indeterminate for AI. These vectors are propelled by post-neoliberal economics and the politics of platform operations and cultures. They also gain strength from the knowledge formations that have enabled ML to multiply and distribute, namely, the cross-pollination of statistical methods, techniques, and its mathesis with computing. Let me turn to another example of how a limited imagining and organizing of the ordering schemas of ML-based AIs sustains this propulsion along closed, autopoietic vectors. It is important to pay attention to how deep learning itself conceptualizes and indeed produces an "aesthetics" through its own computational vision processes and techniques. Here we can learn much about how experience is increasingly machinically cocreated yet also constrained sociotechnically by platform assemblages and by a statistical episteme. Let's look more closely at some examples of ML's autopoietic vector drawn both from everyday engagement with ML AI tools and platforms and from more specialized data science. Data science has increasingly turned to aesthetics as a field for training and testing the autonomous "creativity" of computational machines. It is worthwhile following how this aesthetic turn twists, since it is by working with the twists, obscured by the prominence of the more dominant autopoietic vector, that we might register the differencing of ML.

Developers from and collaborators with platform research and development branches of, for example, Nvidia increasingly use deep learning to "understand aesthetics" (Shaji 2016). The goal of such image-related research is to develop methods of image classification that will automate, for example, the commodified distribution of either stock photographic imagery or select (perhaps even generate) highly aestheticized images to be used, largely, in advertising. This AI-driven "understanding of image aesthetics" is subtended by the datasets of the actual images being classified. In other words, the images must fulfill the functions of both being exemplary as aesthetic images *and* being the substrate out of which a commodifiable machine aesthetics is to be constituted. It is unsurprising, then, that the aesthetics being "understood" here results in a bizarre mash-up of feature extraction and formalism, in which image content is subjected to layers of automated discriminant analysis. Discriminant analysis is one of the oldest and most used techniques of ML, making its way initially into computation from the statistical endeavors of R. A. Fisher's 1936 attempts to discriminate species of irises in larger populations of the flower via measurement values of their leaves and sepals. It is not insignificant that discriminant analysis and its historical genesis continue to permeate methods of contemporary image recognition and generation in deep learning models. Linear discriminant analysis (LDA) is a method of feature extraction from data that aims to find new features that maximize separation between classes or *discriminants*, as they are known in statistics (Alpaydin 2016, 75). As I will argue in chapter 2, ML is indebted to techniques such as LDA, which also conjoined with another social machine—racist eugenics—and complicates the statistical mathesis permeating ML-based computation.

Used at various points and across many kinds of model architectures such as convolutional neural networks (CNNs), discriminant analysis is an approach to pattern detection and recognition in which low-level features of an image such as edge contrast are selected at initial layers of the model, and from these features, increasing details in the image come to be extracted or built. CNNs have been a widespread deep learning approach to processing image data and operate by detecting localized "motifs" in a data input through one set of layers—the convolutional ones. The CNN then pools these motifs and correlates them across the image at another set of "pooling" layers (LeCun, Bengio, and Hinton 2015, 439). "Image" is here conceived as the outcome of layering or building compositional blocks (of pixels, then clusters, then motifs) from local to general, with each block hierarchically subjected to an ascending order of classes from low-level discrimination to higher-level recognition of detailed features: "In images, local combinations of edges form motifs, motifs assemble into parts, and parts form objects" (439). As motifs are detected and assembled across an image instance, and then across other images in the dataset, something more general begins to appear for the CNN: image *style*. Style is defined in the deep learning literature as "a feature space originally designed to capture texture information. . . . It consists of the correlations between the different filter responses over the spatial extent of the feature maps. . . . By including the feature correlations of multiple layers, we obtain a stationary, multi-scale representation of the input image, which captures its texture information but not the global arrangement" (Gatys, Ecker, and Bethge 2015). Understanding how a CNN "sees" image data begins to give us a better sense of why the dominant visual cultural sensibility of ML is located in phenomena such as "style transfer." Having identified "style" as what can be discriminated and extracted from an image or image dataset regardless of content (or, indeed, situatedness, affectivities, materialities, immanent sensory relations, exten-

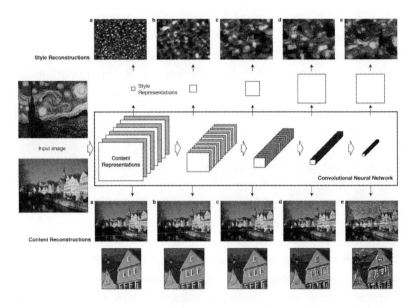

1.4 Style transfer schematically in action in a CNN. Image from Gatys, Ecker, and Bethge 2015.

sive relations with other images or any other contemporary visual cultural or art theoretical criteria), the obsessive goal of standardized ML aesthetics has been to replicate image *style*.

Style transfer transpires as just that capacity of ML aesthetics to discover localized motifs as a new kind of textural "meta-space" that can then be reassigned to any new image instance. An AI model's computer vision that learns to function via the operations of a CNN forms a separate set of spatial and temporal coordinates that are invariant and multiscalar, generating a process for seeding globally applicable "texturality" in the visual field. In figure 1.4, we see how convolution treats the *semantic* content of images as successively expendable as it increasingly passes the image instances across more and more layers of the neural network. What the AI's textural scanning is looking for are initially features in the new image inputs—localized and repeatable motifs—that can correlate with the localized feature maps it has already learned, the CNN's kernel (Goodfellow, Bengio, and Courville 2016, 322–23). A primary operation in CNNs is the training of hidden layers of the network to function as "image kernels," or sets of weights already learned on a training dataset. Typically, the application of a matrix of (trained) weights as kernel to a new input image extracts some aspect from that image and then downsamples it. At initial layers of the network, this might simply be

Pop-Up Definition: Convolutional Neural Networks (CNNs)
Convolutional neural networks (CNNs) are image deep learning models. In mathematics, a convolution is an operation where two functions combine to produce a third. This third function is an expression of the ways in which the first two have modified each other. In a CNN, the pixels of an image input are transduced into a mathematical matrix representing each of the pixels: in a 6 × 6 pixel input, there would be a corresponding array of 36 numerical values. The "convolution" performed on each of these inputs is carried out at the initial layer of the model by what is known as filters or kernels. This is a smaller array whose values act as a filter, passing over the input arrays; detecting, extracting, and convolving features (similar values); and eventually constructing a new array or feature map. The feature map is essentially an expression of the filter having convolved the input arrays. These feature maps or arrays are then passed as new inputs to the pooling layers of the model. Here, further filters are applied, which, for example, extract maximum or average values across the feature maps. The results of pooling are passed to flattened and final layers, which perform classifications or generative image tasks.
Convolutional neural networks began development as models for recognizing handwritten numbers in the late 1990s. They gained widespread recognition in the data science community after the success of AlexNet, a CNN, won the ImageNet visual recognition challenge in 2012. All remaining ImageNet challenge competitions were won by different kinds of CNNs, and these dominated ML image recognition and generation up until 2021 and the advent of text-to-image models. They also gained widespread recognition through phenomena such as style transfer.

an edge; at much later layers in the network, higher-order representations are semantically detected by using object classification layers.

In style transfer, the CNN is also performing a process of correlation, which occurs across all the layers of the neural network, forming a new kind of abstract, nonextensive, and topological space: "This feature space is built on top of the filter responses in each layer of the network. It consists of the correlations between the different filter responses over the spatial extent of the feature maps" (Gatys et al. 2017). Gatys and colleagues' metaphorical use of the phrase "on top of" in the paper that introduced style transfer to the data science community gives us the idea of texture detection as a kind of meta-spatial formative process the AI builds to create a new representative style. In figure 1.4, the CNN initially appears to apply the texture, which its parameters have learned by training on the great masters' paintings, such as Van Gogh's *The Starry Night* (1889). The kernel of creativity is cracked by morphing a picturesque image of Tübingen (the German town in which the research was authored) into textural capture as a feature map of its stylistic intelligence and creativity.[9] In other words, the new image of the Tübingen canal is convolved through the CNN's architecture to feature a *map* of the texture of Van Gogh.

The feature-mapped outputs of ML style transfers such as the Van Gogh Tübingen canal scene look very similar to the ways in which, for example, a software package such as Adobe's Photoshop applies a filter to create a pointillist-looking new image. But the techno-aesthetic processes and the operativity of a CNN and a Photoshop image filter are worlds apart. This is an important point to make yet not altogether easy to see; it requires us to really stare the CNN AI down processually rather than to evaluate it at (sur)face value, where all that is taken into account is an image's artifactuality. What I am proposing follows a very different techno-aesthetic mode of questioning and thinking with ML than, for example, Lev Manovich's "AI aesthetics" (2018). Manovich is concerned primarily with the ongoing automation of aesthetic creation and choice. Regardless of whether this takes place in a Photoshop filter, on Instagram, or through a trained AI model, the issue, for Manovich, is the ubiquity of what he calls industrial-scale "cultural AI" and its overall tendency toward full aesthetic automation (2018, location 13).[10] Within this social trajectory, Manovich asks whether automation will lead to more or less aesthetic diversity. However, diversity here only applies to the products or artifacts of those semi- or fully automated decisions: "But AI, algorithms, and user interfaces of digital services, apps, and products may also be increasing aesthetic diversity. For example, digital cameras and photo apps have many functions for customization. On my camera (Fuji E-3), I can choose the shutter speed, aperture, ISO, desired levels of highlight and shadow tone, color density, sharpening, grain, dynamic range, noise reduction, and film simulation filters" (Manovich 2018, location 93). If we shift from the product and cultural and aesthetic effects of AI to the singular processes at work in ML, then we can see that tensions between sameness and difference are already immanent to the dynamic schemas organizing ML's technical elements. It is not a foregone conclusion that AI as a *technical ensemble* will be responsible for greater or lesser aesthetic diversity. Diversity does not result either from full-scale automation or from increased consumer choice and customization—the twin poles that Manovich suggests organize the contemporary visual cultural landscape. Deep learning enactively produces intensive and extensive sociotechnical multiplicities as it operates. How these multiply to enable sameness or a heteropoietics (aesthetic diversity) is a function of the relations operating with, through, and across the *agencement*.

In the CNN, the issue of how the texturality is generated qua *feature space* is pivotal both to style, experienced in and through ML, and to style's *transfer*—the way a certain field of visuality contoured by computational vision moves into contemporary experience. Here computational processes are transduced by a form of machinic perception (itself increasingly dependent on ML) to produce a particular form of aesthetic homogeneity peculiar to experiments with style via neural architectures. Convolutional neural networks are models with a keen appetite for image *content*, but only that content in image inputs that matches the features already sought after by the model itself: "We can directly visualize the information each layer contains about the input image by reconstructing the image only from the feature maps in that layer" (Gatys, Ecker, and Bethge, 2015). The image content— its semantic information, which largely maps to objects recognized by a classifier layer of the model—extracted by a CNN for the purposes of style transfer is largely drawn from higher-level layers using a classifying function such as VGG-16. Deep learning's visuality is temporally overlaid so as to become correlative with its computational logic: it learns to see what it has produced for itself to be seen at the top level of layers through which content features have convolved. This is nowhere more ubiquitous than in style transfer but also lies at the algorithmic core of what is more generally called computer vision, styled by the dominant paradigm for ML as "representation learning." The latter equates to the best correlation of what is being "outputted" to a schematization or representation of what has originally been inputted.[11] For example, a face recognized—the model's output assigned the highest matching rate—must be matched to a series of computed correlations across the dataset of incoming face image inputs, which, in the deep learning operation of the model, have increasingly been built up as a representation (of a face across the entire dataset). The insistence on ML computational vision as a mode of learning and building representations constrains AI's image generation and visual culture to a logic of equivalence: output = schematized input.

The "style" learned by a CNN, however, works somewhat differently. Style amounts to extracting localized motifs of edges, contrast, brightness, and saturation, which coalesce into a generalized texture when correlated across the many layers across which similar features activate: "To obtain a representation of the *style* of an input image, we use a feature space originally designed to capture texture information.... By including the feature correlations of multiple layers, we obtain a stationary, multi-scale representation of the input image, which captures its texture information but not the global arrangement" (Gatys, Ecker, and Bethge 2015). *Style transfer*, it would appear, occurs when two processes of extraction—content and texture—are distinctly learned by a CNN, which facilitates the combination of different semantic image content, such as the Tübingen canal, with textural information from the master of postimpressionism. However, as Naja Grundtmann argues (2022), the neat distinction between style and content is a human aesthetic distinction and does not map neatly onto the machinic perception at work in ML. Grundtmann demonstrates that, on close inspection, the *style* transferred to the Tübingen canal includes not just Van Gogh's texture but also figurative elements from *The Starry Night* painting. The stars and swooshes of Van Gogh's night sky transfer to the sky above the canal. For Van Gogh's postimpressionist painting, *style* is itself also figurative; he creates figurative elements via techniques such as hatching (close parallel strokes) and color contrast. What is *transferred* here, then, is a style-content relation that is immanent to a particular painterly aesthetics.[12] However, what is relayed in the ML literature about and research experiments with *style* is that this is something that can be computationally learned and re-outputted to restyle something entirely different in time and space.

The Oddness of ML's Space-Times

Equivalence stabilized computationally is the bedrock for prediction. And the key selling point of all ML is its capacity to predict. From the rise and fall of the stock market (Chong, Han, and Park 2017) to claims of accurate forecasting of patient deaths in hospitals (Rajkomar et al. 2018), ML colonizes futurity with increasing probability scores that diminish error rates to below 0.1 decimal points. As a number of people critically engaged with computational culture have argued, the ubiquity of statistical prediction, supported and disseminated by ML assemblages, has much broader sociotechnical consequences, perhaps leading to "a *generalization of prediction* of utterances and actions that progressively interpolates and interpellates subjects" (Mackenzie 2015, 431). However, the actual movements and operations of neural networks—for example, passing inputs through layers, propagating and adjusting weights and biases, pooling and correlating across layers, or, in architectures such as generative adversarial networks (GANs), combatively staging networks that counter each other's outputs—are anything but sequentially progressive. Indeed, a machine-learned predicted output is, generally speaking, correlative with its *past*—it "matches" the information through which it has *become*. This is precisely why data scientists and social and cultural theorists alike argue that a model—and its predictive capacities—is only as good as the data on which it has train*ed*.[13] But here we encounter the rather odd space-time of ML's autopoiesis. We could say that while the generalized program of ML is future oriented, segmenting present data so that it aligns with definitive predictions, the becoming or heterogenesis of ML loops backward to the future, recursively iterating its outputs via what its inputs will have become.

Surveying some of the computational movements at work in AI architectures suggests that the temporalities of ML are at the least multilinear and recursive. A model training—and even a trained model operating on new data inputs—inhabits a space-time that is neither wholly predictive nor sequentially causal, despite these being the twin poles of the orientation of most contemporary ML endeavors. At a basic technical level, the processing of data values occurring in large-scale ML enterprises such as DeepMind's AlphaGo (Silver et al. 2016), an AI that outplayed humans in Go from 2016 to 2023, does not operate serially or sequentially.[14] Rather, most current deep learning models take advantage of graphics or tensor processing units (GPUs and TPUs) that facilitate parallel processing of data. Put simply, these units can contain thousands of computer cores, the computational element that both runs and enacts algorithmic functions. Here data values are split and processed in an independent *parallel* manner across multiple cores and then merged again, or large amounts of the values are simultaneously processed in the same manner.[15] Hence the microtemporalities of ML can generate many simultaneous spatiotemporal realities.

However, the objective in most ML is not to engender differences across multiple events of processing but rather to quickly and efficiently throughput values to fulfill some task: image recognition, text prediction, and so on. To this end, various adjustments to the hardware architecture of GPUs and CPUs have been made, providing microcomputational spaces that contain more "width" to register and process data. We have become familiar with these areas of computation through concepts such as 32-bit architecture, which can hold 32 bits of data at once. In a single-value processor such as an ordinary central processing unit (CPU), each bit takes half of its value to process in one step; that is, for instance, it takes 16 *timesteps* to process

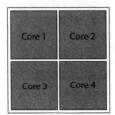
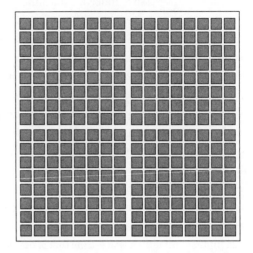

1.5 Schematization of the core functionality differences between a central processing unit (CPU) and a graphics processing unit (GPU).

32-bit data. But the recent history of ML development via platforms turning over large quantities of data has supported the development of *vector* processing, which relies on a combination of greater-width architectures to receive and process data *and* different kinds of functions such as linear algebra libraries (Mackenzie 2017, 69). The greater width means, for example, that a 128-vector processor needs only 4 rather than 16 timesteps to process 32-bit data. Here the quantificatory directions pursued at the level of microprocessing directly affect both the speeds of computing and computation's (previous) sequential logic. Machine learning's great "big data" successes are only possible by distributing its tasks across parallel processing units, which, in turn, signal a shift away from the previous temporal sequentialism toward the parallelism of distributed computational events.

Yet vectorization and parallelism are also important operations and architectures because they *qualitatively engender* potentials for computational novelty through the ways they configure, for example, new microspatialities such as latent spaces. On the one hand, this takes place via the platformization of ML concerned with large-scale and fast task-oriented learning for AI.[16] To this, we might also add the other vectors of quantification that I have been touching on, such as accelerations in the speed of data processing

> ***Pop-Up* Definition: Graphics Processing Units (GPUs)**
>
> **Graphics processing units (GPUs)** are a specialized type of electronic hardware circuitry, whose architecture features thousands of identical computing cores connected in a grid on a single chip. These chips were deployed extensively through the 1990s to accelerate the high-resolution rendering of graphics in computer gaming. They started to be used for general-purpose computing between 2000 and 2010 and increasingly replaced central processing units (CPUs) for deep learning tasks. While CPUs can undertake a wide range of tasks, they cannot do them concurrently. Since GPUs can perform parallel operations on multiple sets of data, they are often used for nongraphical tasks for machine learning.
>
> **Graphics processing units** have migrated from gaming computation to deep learning in part because deep learning has increasingly been concerned with image and high-dimensional data since the late 2000s. The deployment of GPUs as the computational hardware infrastructure of deep learning has also "operationalized" images within the technical ensemble of ML-based AI. Image inputs to ML (even randomized noise inputs) are now transduced, observed, and analyzed by the models in massively distributed ways, enabled by the parallel processing of GPUs. While generative or classified image tasks may be the output of such models, the key function of images in ML-based AI is to operate as computationally observable data.

through reductions in the dimensionality of data quantity; the proliferation and customization of hardware processing architectures from GPUs to TPUs now optimized for deep learning; the rise of large-platform formations such as data centers; and the rapid growth of hardware-driven research and development by corporations such as Nvidia.

On the other hand, vectorization, at both the hardware and the software levels of ML, also creates new kinds of nonoptical, intensive topologies that subtend and distribute across a model's operations and processes. As has already been noted, routine ML operations involve reducing the dimensions of the data being inputted to a function or model without significantly losing the features of that data. Various statistical functions at pretraining and training levels have long been used such as PCA and LDA, or functions and models have been developed specifically for neural network architectures such as autoencoders (Goodfellow, Bengio, and Courville 2016, 494). The explosion of image data generated through social media platforms, and importantly through scientific visualization, has prompted many data science efforts to downsample the mass of data inputs. These vectorizing operations identify similarities and differences in or across all data inputs and extract and then map them as features held relationally together in new "spaces." As we have seen, ML understands such spaces as *latent*, signifying that they are hidden.[17] But *hidden* here does not imply that such spaces can be "revealed." In simpler dimension reduction processes such as PCAs' use in early attempts to recognize facial features, a visual artifact presenting the latent feature space could be generated.[18] But in models with many layers and millions of parameters, the latent space is distributed across the model as it runs, as we see with text-to-image models. Here latent space cannot be visualized or represented as such, since it exists only in the model's processes. Elsewhere in deep learning architectures, latent spaces do not actualize during a model's operations but come into being *postoperationally* as spaces to be further analyzed: "We can consider a generative model . . . where we assume that there is a set of latent factors that generate these multiple views in parallel, and from the observed views we can go back to that latent space and do classification there" (Alpaydin 2014, 511). Here we are invited to enter a space that in some ways only exists retroactively.

Latent spaces in ML are space-times of potentiality, which may actualize in different ways: they can be activated to produce visual artifacts; they might be mobilized as distributive vectors as a model trains or runs; they might be invented post hoc as what was retrospectively required by the model to function and which is then *construed by further operations on a model's "views."* Here the actual technics of ML are performed in its very operativity. In the first two cases, latent spaces are operative in or during the generation of images and visual instances but do not lend themselves to being "revealed" as such, in spite of a growing will within the data science community to visualize a model's hidden processes (see, e.g., Yosinski et al. 2015). Instead, they are diagrammatic in the sense conjured by Charles Sanders Peirce; that is, they are schemas that, through their relationality, engender conditions for producing something new out of that relationality (Peirce 1933, 531; 1998, 303). As potentialities that drive the operativity of ML, latent spaces afford ML immanent indeterminacy, or we could say, returning us to a Simondonian terminological genealogy, latent spaces signal machine learning's margin(s) of indeterminacy.

For Simondon, indeterminacy occurs in marginal ways for a technical object, but the ongoing development of an object—especially as it becomes something like a technical ensemble—*positively incorporates these marginal elements into aspects of its own functioning*: "In the concrete technical object, all the functions fulfilled by the structure are positive, essential, and integrated in to the functioning as a whole; the marginal consequences of the functioning, eliminated or attenuated in the abstract technical object by corrective measures become stages or positive aspects in the concrete object; the schema of functioning incorporate marginal aspects; consequences that were irrelevant or harmful become chain-links in its functioning" (2017a, 39). We could even say that latent space—once simply a means of compressing large amounts of data—now becomes central to the training and operativity of models such as text-to-image deep learning networks. What was once an inconsequential operation on quantity has become something qualitatively core to the potential for different kinds of spatiotemporal events in ML computation.

Fooling Computational Vision

While style transfer as an aesthetic epiphenomenon of convolutional neural networking has been aesthetically ascendant in computer vision first through CNNs and then in a more dispersed means via the dominance of photorealism and 3D CGI styles of text-to-image generators, this is not all that an aesthesia of deep learning offers.[19] I want to change gear now and focus on the ways in which the eventfulness of ML itself can become a site for experimentation within data science itself. We do not have to look far to encounter a weirdness amid machine learners, an example being the AI bots "Bob" and "Alice," trained by Facebook researchers in adversarial negotiation using English natural language processing GANs (Lewis et al. 2017). At a certain point in their networked negotiations, the AIs began to develop their own dialect "language," a kind of pigeon "English-enumeration":

> Bob I can I I everything else.
>
> Alice Balls have zero to me to me to me to me to me to me to me to me to. (Lewis et al. 2017)

Facebook, ultimately interested in training bots that could communicate *with* humans rather than with each other, shut the model down and, after readjusting how the networks generated and discriminated in their relations between each other, "rewarded" their adherence to human comprehensible English alone (Lewis et al. 2017). What is interesting here is not what the bots

```
Alice : book=(count:3 value:1) hat=(count:2 value:1) ball=(count
Bob   : book=(count:3 value:0) hat=(count:2 value:0) ball=(count
--------------------------------------------------------------
Bob   : i can i i everything else . . . . . . . . . . . . . .
Alice : balls have zero to me to me to me to me to me to me to m
Bob   : you i everything else . . . . . . . . . . . . . . . .
Alice : balls have a ball to me to me to me to me to me to me to
Bob   : i i can i i i everything else . . . . . . . . . . . .
Alice : balls have a ball to me to me to me to me to me to me to
Bob   : i . . . . . . . . . . . . . . . . . . . . . . . . . .
Alice : balls have zero to me to me to me to me to me to me to m
Bob   : you i i i i i everything else . . . . . . . . . . . .
Alice : balls have 0 to me to me to me to me to me to me to me t
Bob   : you i i i everything else . . . . . . . . . . . . . .
Alice : balls have zero to me to me to me to me to me to me to
```

1.6 Screenshot of the Facebook adversarial bots, Bob and Alice, in text exchange.

really meant or how they may have erred or strayed. Rather, in the process of trying to produce communicational exchange, the model itself begins to create its own redundancies through multiplying elements of its lexicon and syntax. Its redundancies or "errors" become *efficient* for the purposes of sustaining exchange between them. What we see with Bob and Alice in the exchange that iterates in figure 1.6 is not *human* language but instead the production of conditions for machine *communicability* through multiplying the potential departures and arrivals of conversational threads. This is not signification per se but an elaboration of a field of asignifying potentialities from which communication might emerge: "a-signifying semiotics which, regardless of the quantity of significations they convey, handle figures of expression that might be qualified as 'non-human' (such as equations and plans which enunciate the machine and make it act in a diagrammatic capacity on technical and experimental apparatuses)" (Guattari 1995, 36). Rather than construe the bots' departure from English as "error laden," as was Facebook's reaction, it might ultimately be more interesting to allow AIs to perpetuate these kinds of operations. These suggest that something "more than" an instrumental operativity both conditions and exceeds a fixed task to be achieved such as natural language processing or image representation. Bob and Alice are less bots that are trying to reproduce human capacities than enactments of the multiplicatory phenomena that make up computational experience. Affirming their asignifying operativity is not the same as cultivating the "erroneous" within AI. I am not advocating for an extension of "glitch aesthetics"—as has been common in attempts to create alterities for digital media (e.g., Nunes 2011)—to ML experience. Instead, I am suggesting we affirm the capacities of the *agencement* of ML to actualize according to its processual tendencies. Many other instances of these *machinic asignifying tendencies* are emerging throughout various scientific and platform deployments of machine learning.

In a study that initiated a new strand of pattern recognition research and provoked much debate about computer vision (Nguyen, Yosinski, and Clune 2015), data scientists generated synthetic images using evolutionary algorithms in conjunction with a deep neural network classifier. The task for the AI here was to classify these synthetic images to one thousand or so classes—such as lions, motorcycles, digits, and so on—for which the model had already been robustly trained. The AI used the well-known Alex-Net architecture, a CNN trained on the 1.3-million-plus dataset of images, ImageNet. At the time, AlexNet was a benchmark of deep learning success for computer vision (Alom et al. 2018). Both in laboratory settings and on platforms, AlexNet's architecture has been and is still used to implement aspects of image recognition within complex machine learning platforms such as Google's TensorFlow. But in Nguyen and colleagues' study, in more than five thousand runs of AlexNet on images they had synthesized using evolutionary algorithms, the AI consistently mislabeled images. The images, according to these computer scientists, were "fooling" ML's best computer vision operative (Nguyen, Yosinski, and Clune 2015).

Yet this was not a case of "error" but instead one of the model performing *inconsistently via its very functionality*. Importantly AlexNet is *discriminative*, which, as mentioned earlier in this chapter, involves initially recognizing data inputs through low-level feature detection such as edge contrast—discriminating between light and dark—in images. Increasing details in the image are built from these initial feature detections, so that eventually an input is matched to a particular class: lions, motorcycles, starfish. In images drawn from the study (fig. 1.7), the image labeled "starfish" contains the blue of water and the orange of a starfish, which the AI would already have learned from the original ImageNet dataset on which the model was trained. The discriminant analysis detects the presence of blue next to orange, then, in the synthetic images being inputted as part of the inputted image's "global picture": "For many of the produced images, one can begin to identify why the DNN believes the image is of that class once given the class label. This is because evolution need only to produce features that are unique

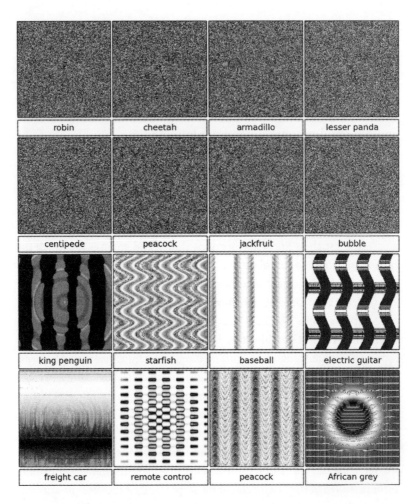

1.7 Images drawn from Nguyen, Yosinski, and Clune 2015, accompanied by a caption that reads: "Evolved images that are unrecognizable to humans, but that state-of-the-art DNNs trained on ImageNet believe with >= 99.6% certainty to be a familiar object."

to, or *discriminative* for, a class, rather than produce an image that contains all of the typical features of a class" (Nguyen, Yosinski, and Clune 2015, 5).

The model labels or classifies incorrectly here *because* it has learned to do what it is supposed to do: discriminate. It is possible to speculate that the new synthesized image inputs pose a new "class" that the model has not yet encountered, and so it resorts to recognizing these images in the best way that it can. But even if this is the case, a functioning, robust, and ac-

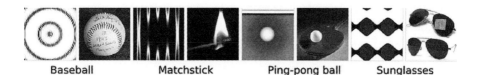

| Baseball | Matchstick | Ping-pong ball | Sunglasses |

1.8 A selection of paired images that show the synthetic image and its target class match submitted by Nguyen et al. (2015) to the 40th Annual Juried Student Art Competition, University of Wyoming, 2015.

curate model nonetheless reveals either that it accurately *mis*labels or that, given a kind of data input, its classes, and hence its syntax of classification, become radically contingent. Either way, we have a model processing in a heterogeneous manner. Here we might concur with Parisi's analysis of the differentiating tendencies within parametric computing (2013, 86). Variability is added from "outside"—in this instance, the synthetic images are the variables—and is intensively produced, immanently generating as the parameters, or here the "weights" and "biases," come into relation with inputs as the model runs. Relations between data and systems such as neural networks do not have to adhere to predictive outcomes but also allow novel forms of computational contingency to emerge.

What Nguyen and his colleagues concluded from their study is that this openness to "accurate" error or to persistent mislabeling might mean that such networks are not generalizable. This is a radically different proposition from much of ML research and its widespread application via platforms, both of which strive for generalizability and optimizing predictive accuracy (Mackenzie 2017, 6).[20] Nguyen and colleagues also went a step further to argue that AI visual perception modeled via deep neural networks is indeed *different from*, rather than similar to, human vision. In a move that hints at what different directions might be available through pursuing the radical contingencies of computational processes, the researchers submitted some of their image results of close pair matches between synthetic and target image classes to an art competition at the University of Wyoming (fig. 1.8). The museum's acceptance of their submission opened something more speculative for a deepaesthetics, named by the researchers as "computational creativity" (Lehman et al. 2018). Rather than seeing the neural networks as limited or erroneous, they noted the possibility of "combining the discriminant deep learning networks with evolutionary algorithms [for] more open-ended, creative search algorithms" as a potential outcome

(Nguyen, Yosinski, and Clune 2015, 8). In acknowledging the potential for open-endedness in these category "mistakes," a sensibility of contingency and indeterminacy can be cultivated.

Computational Creativity or Creative Computation?

The question of creativity's autonomy—does it belong to the human alone, or could an AI be just as creative?—is revived each time ML-driven computer vision scales up, not only technically but also because of platforms' economic and cultural investments. In 2012, Google's Brain, a network of sixteen thousand microprocessors, learned to recognize cat faces in videos without first being shown images that had been labeled as cats. This intersected with the explosion of cat videos on YouTube and also spawned a range of online cat generators, including Alexia Jolicoeur-Martineau's Meow Generator, which synthesized more than fifteen thousand nonexistent and sometimes creepy cat faces (Jolicoeur-Martineau 2017). Google Brain's breakthrough in 2012 was notable both for its capacity to train a network for grasping high-level representations such as faces and because this was made possible by platform computational services and hardware infrastructure. We have already seen that Google's DeepDream also spawned a particular style of pareidolic imagery and ignited a debate about the unconscious creativity of AIs. In the 2020s, we see computational creativity being given another push as AIs generate images that resemble high-resolution computer graphics by being simply prompted by a string of text. As data scientists note (Goodfellow, Bengio, and Courville 2016, 440), much AI computer vision is dedicated to replicating human capacities. Where computational novelty or creativity is asserted for a model, such as in a creative adversarial network (CAN)—an attempt by deep learning researchers to specify a more general generative adversarial network of GAN architecture toward producing creative (image) outputs—*human creativity* is, nonetheless, the comparative benchmark: "We conducted experiments to compare the response of human subjects to the generated art with their response to art created by artists. The results show that human subjects could not distinguish art generated by the proposed system from art generated by contemporary artists and shown in top art fairs" (Elgammal et al. 2017, 1).

Despite this, the actual aesthetic and technical delimitations of the CAN are worth examining more closely because its modelers sought to encode the assemblage with both determinate and indeterminate capacities. Here, then, we see a struggle of forces playing out within ML between the predictive and the contingent. The researchers behind the CAN describe this as a kind of battle between emulating and deviating from style transfer: "We focus on building an agent that tries to increase the *stylistic ambiguity* and deviations from style norms, while at the same time, avoiding moving too far away from what is accepted as art" (Elgammal et al. 2017, 3–4). This can be reframed as an attempt to produce a model at the nexus of continuous and discontinuous vectors. In this reframing, I think we can reposition experiments such as the CAN as offering a glimmer of a sensibility emanating from data science that is, in spite of itself, heteropoietic. This glimmer, however, sits within an entire corralling of art and aesthetics by ML more generally into an art historical taxonomy, delimited by clearly categorized and recognized styles, movements, and media. The operations and outcomes of such taxonomizing are part of a much closer alignment between information science, digitization, and the building of art databases between museums and galleries since at least the mid-twentieth century. This alignment plays out in, for example, Robert Glushko's (2016) text, addressed in part to museum professionals involved in resource description. Glushko asserts the organizational principles of interpretability in art historical codifications of classification found in, and recommended by, Erwin Panofsky's iconology that, Glushko argues, support human-legible resource classification of digitized collections and databases. The crisscrossing of information science and art history in the organization of databases and resources for libraries and museums folds back into the training of models via, for example, the use of the WikiArt dataset (2010–) by the CAN.[21]

In spite of tying art to a broader sociotechnical classificatory schema, the CAN hints at the production of novel experience via ML. To see this, we need to think its technicity and aesthetics processually, at the level of the processes and operations it enacts. The CAN piggybacks on the network architecture of generative adversarial models (GANs). The basis for this architecture lies in the two different networks set up to compete against each other as adversaries. One network is termed a *generator*, with the task of generating artificial outputs of, for instance, images or handwriting samples. Over many iterations of producing these outputs, the generator becomes "good" at synthetically reproducing the same distribution of variability across such outputs. The other network, the *discriminator*, has (usually) learned this distribution by training on a given dataset. Such datasets might include actual images of people's faces; for instance, the Flickr-Faces-HQ (FFHQ) contains seventy thousand high-resolution human faces scraped from the Flickr platform and was originally created as a benchmark for GANs.[22] The

> ***Pop-Up* Definition: Generative Adversarial Networks (GANs)**
>
> A **generative adversarial network (GAN)** is a composite of two neural networks: the generator, which creates new synthetic outputs, and a discriminator, which classifies these outputs. The two networks are mathematically set to play out an adversarial "zero-sum" game in which one network gains and the other loses. The discriminator "discriminates" on the basis of how proximate new instances being furnished by the generator are or are not to a learned pattern of feature distribution. Similarities result in a positive output signal sent to the generator. The generator then adjusts its weights toward the creation of new data instances that increasingly become more proximate to the distributions of the discriminator. As this process continues, a state of equilibrium or zero-sum game is eventually arrived at where the generator is producing instances that no longer differ from the feature distributions held by the discriminator.
>
> GAN architectures were first introduced in 2014 and quickly became the neural network basis for synthetic image generation, especially after Nvidia's release of StyleGAN in 2018 with its capacity to create new "fake" human faces. These networks have been widely used by artists, often due to their inherent architectural instability (the two networks can cause each other to crash), and for the visual access they provide to their latent spaces.

discriminator literally discriminates the generator's outputs on the basis of their adherence to the distribution pattern, rejecting some outputs that do not adhere and reinforcing the generator's outputs that do fall within the pattern. Eventually, the generator is trained to output, for example, synthetic face images that look as if they belong to the original FFHQ dataset. When the GAN begins initially producing image data instances, they often look like varying distributions of random grayscale noise. Only some of these distributions will have early proximity to the distribution of features being trained for. It is the task of the discriminator—a classifier model—to compare feature distribution in the instances produced by the generator to the distribution in instances it has assimilated as weighted parameters in its neural architecture from the original training dataset.[23]

It is not surprising—given the endgame is one of equilibrium of features between new instances and training dataset instances—that GAN architectures have also increasingly been used to emulate the style (understood as a mean distribution of certain features in the data) detected in image datasets.

Such features, in the case of models seeking style, correlate to certain distributions of, for example, edges, lines, strokes, and so on, detected as textural patterns across the original dataset of images. This has led to GANs generating new instances of recognizable imagistic styles, including photorealism. Researchers at Nvidia, for example, used an image dataset of celebrities to train a GAN to produce photorealistic faces of synthetic celebrities (Karras et al. 2018). The results are uncanny in that they are both immanently celebrity-like yet not ascribable to any actual human celebrities. Since there is no longer any difference *for the model* between the actual celebrity images and the artificial nonexistent celebrities, we might ascribe these steadfastly realistic yet nonindexical images the status of "real fakes." Indeed, the terms often used by deep learning researchers in describing the work of GANs lend themselves to such descriptions, for example: "This drives the discriminator to attempt to learn to correctly classify samples as real or fake. Simultaneously, the generator attempts to fool the classifier into believing its samples are real" (Goodfellow, Bengio, and Courville 2016, 697). The distinction between real and fake is something of a red herring given that all data instances used in deep learning architectures require some mode of computational optimization, inviting the question: What were the real or raw data in the first instance? Nonetheless, the divergence and convergence of real and fake have become powerful criteria for evaluating the aesthetic value of ML image outputs; the more realistic the synthetic fake becomes, the more gracefully or ingeniously the GAN has operated. Here aesthetic value is accorded on the basis of a generalized equivalence, reaching all the way from the ML architecture, where the generator's outputs become equivalent to the discriminator's, to the online attention given to sites such as Thispersondoesnotexist, which instantaneously creates new face image instances of nonexistent people.[24]

In the generation of high-resolution synthetic faces and in other style transfer phenomena, any *heterogenesis* for AI is, however, foreclosed, since the model's very operations move toward convergence and overdetermination. Here, GANs become another mechanism for promulgating visual equivalence underwritten by a model that progressively interpolates novel data instances back into ones already recognized. This is further replicated in the rash of popular online sites such as Picbreeder and Artbreeder emerging on the back of GAN image synthesis research undertaken by Google's DeepMind and DeepDream.[25] Here the claim for GAN aesthetics is the reproduction of a synthetic realism peppered with alluring twists: "Art using neural models produces new images similar to those of natural images, but with weird and intriguing variations" (Hertzmann 2019). Novelty is ground down to featuring as variability within a narrow distribution of the same. But if we look instead at the CAN rather than the computational creativity ascribed to GANs in general, we can detect something else at play.[26]

The aims for the CAN are to get the generator network to synthesize unclassifiable "not-art" (or, as the researchers term it, "stylistic ambiguity") that is nonetheless still recognizable by the other network, the discriminator *as* "art" (Elgammal et al. 2017, 3–4). It is worth noting that the discriminator's recognition of "art" is based on its prior training of standardized art classes or styles derived from the WikiArt dataset of fifteenth- to twentieth-century canonical Western paintings. Yet the CAN's adversarial tension between its two networks is not set to resolve to a zero-sum equilibrium in which the generator's synthetic instances must be complete matches to the discriminator's prior training. Instead, the model's neural networks subject classification itself to a process of ongoing differencing where novel instances—here images—are "seen" by the AI as both not-art (in the sense of art styles not recognized by the model's previous training) *and* art simultaneously. No longer category mistakes, the CAN's flow of generative art is a GAN architecture operating in a mode similar to the "both-and" hybrid image that resulted from Stable Diffusion's wrestling with the rabbit-duck illusion (fig. 1.2). To keep the AI producing at this conjunction of discontinuity and continuity, the generator cannot simply converge with what the discriminator has been trained on, as is the case for most GANs. In addition to the numerical values generated by the discriminator detecting stylistic proximity in an image produced by the generator, in the CAN the discriminator also produces another value indicating whether the style being generated is too easily classifiable according to the categories the discriminator has learned from the WikiArt dataset. The new synthesized images are the product, then, of an oscillation between an affinity to the styles detected in the WikiArt dataset and a rejection of strict adherence to those styles. The overall goal for the CAN—now a complex ensemble of ongoing interrelations between negative and positive values—is to produce new image instances that continue to hold open the possibility that they are both similar to the art they trained on and yet not recognizable as that art at all. Unlike Nvidia's synthetic celebrity faces, in which convergence toward detectable photorealistic likeness remains key, the CAN holds open an indeterminacy between the new synthetic images generated and their canonical likenesses.

Here we might begin to tease out another deepaesthetics, which engages directly with questions of process, indeterminacy, and difference rather than prediction. What the oscillatory signaling of the discriminator in the CAN both discloses and occasions is a machine (learning) dynamic whose synergy is novel at the level of process, resulting in what Andrew Murphie terms "a kind of moving flow of difference" (2019, 32). Whitehead's and James's conceptions of creation and aesthetic feeling resonate here as "novelty" individuated through variation while maintaining a relation to identity, sameness, or, as they express it, continuity. Discontinuities emerge within (are immanent to) continuities in such processual approaches to experience. For Whitehead, "all aesthetic experience is feeling arising out of the realization of contrast under identity" (1978, 280). For James, a distinct "feeling" is what qualitatively contrasts with its adjacencies but is only experienced via the relations of transition or continuity across qualitative contrasts (1890, 249). In Whitehead's articulation of aesthetic fact, the oscillatory functioning of the CAN's contrasting yet conserving operations resounds:

> An intense experience is an aesthetic fact, and . . . its categoreal conditions are to be generalized from aesthetic laws in particular arts.
> The categoreal conditions . . . can be summarized thus:
> 1   The novel consequent must be graded in relevance so as to preserve some identity of character with the ground.
> 2   The novel consequent must be graded in relevance so as to preserve some contrast with the ground in respect to that same identity of character. (Whitehead 1978, 279)

Whitehead's "novel consequent" at once carries some sense of a new instance emerging (or what he would call an actual occasion) and conserves its relations to an ongoing "ground." Considered not so much as isolated instances or moments that arise, aesthetic "feeling" as experience—a conjoining of Whitehead and James—could help us register an aesthesia specific to a time of ML, a time in which a moving continuity or flow of differencing comes to qualitatively shape the creativity of and creative encounters with AI(s).

Perhaps, then, the CAN is a less a machine for producing artworks as instances of (human) creativity, as its modelers claim, and more a machine for learning to move (with) flows of images that immanently difference. We humans cannot "see" this moving flow despite all the images the CAN does and might continue to produce. Instead, we register the aesthetic feeling of such moving difference both across the relationality of the CAN images (especially when seen as a grid of selections out of the vast output of

1.9   Example of images generated by CAN from Elgammal et al. 2017.

all the model's image instances) and in their differencing relation to their ground—the canon of Western painting. In figure 1.9, a human-selected slice of images made by the CAN, the operations of variation and repetition function both against an extensive ground, the eighty-thousand-plus WikiArt dataset of images, and in relation to an intensive "plane" generated amid the CAN images' processual relations of contrast, motif repetition, and deformation traversing the new synthetic instances.[27] The human modelers think of the CAN's novel images as qualitatively "abstract" (Elgammal et al. 2017, 11). But they turn away from the potential for the model to be generative of something stylistically heterogeneous by seeking validation for the CAN's creativity in evaluations they stage for a human audience to make. The CAN images were favorably compared to selected images of human artworks exhibited at Art Basel 2016 (Elgammal et al. 2017, 13–21), which returns the AI to an aesthetic economy of generalized human equivalence.

But the resemblance of CAN images to contemporary human paintings is not what is important for an exploration of computational experience. Rather, a (deep) aesthetic feeling—which we might well call "abstract," since it no longer resides in the images themselves—emerges via the variation-repetition visual movement generated by the CAN model as it spits out new image instances. We might register this different (deep) aesthetic feeling as a differenced continuity, or continuity registering as change changing (Massumi 2002, 10); of difference moving a ground, a tradition, a canon of sameness; *modulating it yet retaining consistency*. Such modulation occurs at two levels in the CAN model: first, the CAN is "productive," resulting in actual new data instances or images that occur at a nexus of differencing. The novel images carry traces of both continuity of style and distinct modulations of that style. Second, as a new kind of "abstract" image, these CAN occasions—together with the consistently inconsistent "fooling images" or the asignifying yet communicable syntax of the Alice and Bob bots or the illusion of the rabbit that becomes duck—form a vast and ongoing ensemble of new kinds of images, which become a "pool" of potentiality. For such images, syntax and other data instances will aesthetically, culturally, and computationally feed into future ML models as these scrape image data from wherever they are posted online. The models, then, *set up conditions for self-modulation*.

Registering all of this as an aesthetic feeling specific to the processuality of ML differs from cultivating an aesthetics of, for example, style transfer, which amounts to not much more than seeing variation against an unyielding background of sameness. It also diverges from the claims for an aesthetic peculiar to the discrete character of computation that, for example, Fazi claims must be attributed to the digital (2018b). Fazi has argued that the digital must be taken at the level of its own ontological reality, which is this discreteness: "The digital is a data technology that uses discrete (that is, discontinuous) values to represent, store and manage information" (2018b, 2). This discreteness also extends to the axiomatic or algorithmic activities of computation, which, she argues, proceed via rule-governed operations on this "mass" of discontinuous information in equally separate and quantitative steps (14). However, Fazi (2018a, 113) also proposes, following Turing and Gödel, that computation deals with infinite and incomputable quanta. For her, this introduces the possibility of quantitative indeterminacies native to computation as the ground from which novelty might arise. In other words, Fazi reads Whitehead's aesthetics as the novelty arising from the contrast between finite and infinite quanta playing out within the overall discreteness that characterizes her analysis of the axiomatic and quantitative sides of (digital) computation. But I do not think that an aesthetic ontology for ML can be sought in discreteness, despite it too having infinitely quantifying tendencies. Although so much of ML is oriented to measuring experience, it is also a mode of occurrent learning: the reality that some *difference* is generated *as computation happens*. If we think only of a model itself as an instance of (limited) computation, we locate that change as the experience the model gains across its network by vectorially mapping the *relations* of inputs to outputs. Any kind of ML understood as "a computer [that] is said to learn from experience E" (Mitchell 1997, 2) involves relations in which the algorithmic (or "axiomatic") and data (quantities) are in incessant and entangled dynamic codeterminacy.

Computational experience in a time of machine learning can be considered broadly as change occurring as the result of multiple *relations* generating via ML operations. If there were ever a form of computation that materialized the real of relations, then it would have to be the ensemble that is ML. We humans do not experience the bulk of these relations; we do not *live* them. But that is not to say they are not occurring, registering or conditioning a slew of empirical and abstract domains. Although frequently oriented toward a determinism that predicts outcomes based on large quantities of inputs, most ML nonetheless deploys a *probabilistic* logic. Here there is a *likelihood* of a predicted outcome, but a margin of error is always also operative. Of course, as I have suggested by looking at phenomena such as style transfer—as well as in many other applications of ML, from predictive texting through to facial recognition—the aim of the model is to match the outcome as closely as possible to a (pre)determination. The many recursions of training to which a model is subjected are often attempts to reduce its error rates. Despite this, we might think of this margin of error that must inhabit a probabilistic mathesis as just that very indeterminacy embedded in the calculability of AI models. It is what must facilitate updating, adaptation, and change for learning or indeed for experience to have taken place. And this ongoing calculability of the model requires each new parametrization to retain some trace registration of the previous set of parameters or weights.

In the GAN architecture, each layer pass in the neural network engenders both a collective registration of the previous outputs *and their changes across the layer* as they all pass to the next layer. This series of relational processes can best be seen by looking at what happens at early layers of a GAN model, where random, noise images both ingress their generative noisiness and increasingly become discriminated according to certain low-level features such as edges, lines, and contrast. The GAN's very calculations are predicated on its ongoing capacity both to retain its previous calculations *and* to change them. And herein lies its *processual* continuity, which functions in tandem with its calculable *discreteness*. It is not simply that ML as a mode of computation responds to incomputable numbers. Instead, ML is a mode of computation that performs relation across the disjunction of quanta and qualia. It is here that the contingent and novel might arise *so long as ML also faces toward other cultural, social, political, economic, and art machines that engage and co-compose for such novel potentialities.* Machine learning is not simply discrete calculation or axiomatic execution. It is instead the qualitative machinic transduction of the dimensions of the quantificatory as they come into relation with intensive and extensive sociotechnical vectors. We must insist on ML as *agencement* of the entire field of its actual technics. As we have seen, the intensive vectors of ML can be appended by the different kinds of computational hardware on which ML depends and which it coconstitutes: the GPUs now customized for deep learning frameworks. In turn, these technical elements draw and multiply ML in the direction of operative formations already in train—platforms, for example—and emergent novel concepts and events such as new micro-spatiotemporalities. The challenge for a novel aesthesia in which ML plays a vibrant, rather than reactionary, role is to carefully probe at and think with the sociotechnicality of ML experience as it moves and multiplies its relationality via slippages and transductions of quantity and quality. The category mistakes, fails, fooling images, and nonart are all elements that signal ML's margins are open to (its) ongoing computational contingencies.

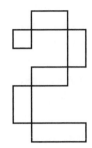

The Color of Statistics

Race as Statistical (In)visuality

It is unusual to perceive statistics and its endless data points, lines, rows, and columns of numbers as colored. In standard histories of statistical graphics, which nowadays fall under the seemingly sexier and far more ubiquitous moniker of data visualization, we can identify two golden periods of an explosion of images representing quantitative information: the second half of the nineteenth century, and from the mid-1970s to the contemporary period of high-tech, high-resolution "data viz" (Friendly 2008, 513, 524). It is certainly the case that colorful images, whether multivariate weather charts, pie charts and bar graphs, scatter plots and heat maps, abound in both these periods, during which statistics and functions, data, and models have been in ascendance. But when we visually represent statistical information as calculable and computable entities, the color fades, and we are reduced to, for example, a black-and-white table of numerical points. These seem to hold their immanent correlations and patterns to themselves, and then we call on "color" *supplementarily* to unlock and visualize them.

In this chapter, I explore a series of lines that unfold from what appears at first to be a whimsical aesthetic question: What if statistics were already deeply colored? If we were to stay only at the level of "hue," we might tour the ways in which the pictorial aspects of the statistical have moved from mono- to polychromatic forms of representation. Yet for anyone alive to the heated debates and discussions on algorithmic and data bias that populate online platforms, and in the pushback from Black, Brown, and many more colored voices within the AI industry, a color line with very different hues than whimsical ones emerges. As Saidiya Hartman has argued, classificatory schemas enact regimes of social and political division between races through which a definitive color line is produced via forms of categorization (1997, 187). This is marked not so much by the production of two equal categories of Black and white peoples but via an epistemo-political securing of privilege via a division between bodies. Here a polis and society—North America, for example—come to subordinate Black bodies and life to a line in which whiteness is the norm. As we shall also see, and as much work on the social and epistemic history of statistics has already revealed (e.g., Hacking 1990, 160; Desrosières 1998, 236), classificatory operations lie at the heart of a certain knowing of the world. My question about statistics' deep "coloring," then, is aimed not simply at how they deploy color in the visual representation of data. I want to consider whether a color *line*—obscured by the black-and-white enumeration of statistics as mathematical and axiomatic—lies at the core of statistics' *mathesis*, a color line that determines its ways of making the world knowable. Statistics' rows of comma-separated values (CSVs), its strings and arrays of transduced data, its tabulation of (data) dimensions, for example, may not be so black-and-white, after all. Their whiteness and blackness signal another kind of color line running through statistics' depths.

This is not to say that uncovering this line—a line that generates and enacts racism and through which statistics and its techniques are historically and contemporaneously entangled—will make it go away. Instead, my pursuit of statistics' color involves a double articulation of ML's *agencement* as at once visually available to encounter at the level of its (black-and-white) data points *and* invisual in the registration of its social and political forms of racialization. Here the concept of the invisual does not mean the invisible or opposite of visual. Instead, it refers to the ways in which visual elements such as images, image collections (datasets), and forms of machine perception such as computer vision have been operationalized beyond their visual functions by the *agencement* of ML. That is to say, the assembling (*agençant* in French) work of platforms, technics, social machines, logistics, and data practices deploys optical elements, modalities, and techniques of observation but uses them for nonrepresentational means and ends. In this sense, the visual is still a register within ML's *agencement*, but one made operative rather than working at the optical/representational plane.[1] It is this operative invisual register, functioning but never seeable as such, through which a racialized color line unfurls. It moves through statistics' genesis to unfold and permeate ML. Its color line is constitutive of statistics' actual technics; it works as a sociotechnical operative to actualize statistics as racist. Rather than only calling out the bias of ML's statistical racism, we need artful techniques that work to redraw color. A processual seeking out of this line's invisuality at the granular level of ML operations is one such artful technique. I pursue this by following the lines, curves, and spaces generated by two figures or functions that begin their life in the historical nexus of statistics and eugenics: principal component analysis (PCA) and linear discriminant analysis (LDA). Later in the chapter, I look in detail at how Stephanie Dinkins's work with AI attends to a different imaginary that she calls the "Afronow." Dinkins's artfulness gives us a coloring of AI in which blackness is no longer subordinated to a normative line of whiteness. This in turn facilitates another kind of deepaesthetics in which statistics' invisual but deeply felt color lines might be unsettled, opening AI to different communities and practices and novel ways of (machine) learning.

To tap into the invisuality of statistics, we will need to mobilize different "ways of seeing" to observe what it already supplies us in its operations and processes.[2] Geoff Cox (2017) has argued that ML, and algorithms generally, cannot be understood to see like a human but instead trouble naturalized Western conceptions of seeing and its seamless connection between the seen and the known. While recognizing the importance of these arguments, my interest in different ways of seeing statistics' numerical visual displays also draws on approaches to visual perception from process philosophy. Here the emphasis has been on the relationality of all perceptual modalities, but particularly how (human) visual perception is conditioned by a field of movement that both cuts across the actions of perceiving/viewing and informs how images come to *image* (Manning 2009, 83ff.; Massumi 2011, 127). If we look at very basic forms of statistical display such as data tables, we can see that more than a simple visuality is shown. Even pared-back displays of data in rows, tables, and columns vibrate with relations in which extra registers activate and complicate their visuality. In turn, this suggests that displays of data are not simply ways of organizing and representing the numerical and statistical to make patterns legible. I want to suggest instead that statistics is already a topology of relational surfaces that can be detected if we register its visual displays—tables, arrays, strings, and so on—as diagrammatic. Here the diagram is less a map of explication and more derivative of a manifold of relations in which data and algorithm/model are always operating coconstitutively.

Data's Relational Opticality

A case in hand is R. A. Fisher's 1936 *Iris* data, drawn up to reduce three species of deep blue-purple iris growing on the Gaspé Peninsula in Canada to a series of measured variables available for comparison (fig. 2.1). Here the irises' deep tints have faded into the monochromatic realm of data points, starkly set in black and white. Fisher's *Iris* dataset is one of the most variously visualized and disseminated in statistics, being bundled with software packages such as R; available with libraries such as Matplotlib in Python, a high-level multipurpose programming language that underscores much current data visualization; and used as the standard test case for many statistical techniques in ML.[3] But the rigorous tabulation in Fisher's 1936 article (although not its actual collection, which was carried out by the botanist Edgar Anderson and given to Fisher when Anderson came to work in his laboratory in Britain) attempts to separate processes of measurement from their subsequent visual representation as chart, plot, and graph.

Or does it? The very tabulation of the *Iris* data effects something peculiarly optical. The typesetting of the diversely increasing and decreasing measurements in columns produces a sensation of swaying movement both down and across the page. Numbers, when organized together in a data structure such as this table, seem to be already information, choreographing the visual field and unsteadying the fixity of their identity as precise points in time. Importantly, then, the data*set* is already more than a collection of data points, measurements, or numbers, but a way of bringing data into relation accomplished by a data structure such as a table. Although outwardly colorless, the data's relations visually manifest for us choreographically in the way the table lays out directionality for the points horizontally, vertically, and diagonally. A dizzying vectoriality slices through and across the *x*,

Table I

Iris setosa				Iris versicolor				Iris virginica			
Sepal length	Sepal width	Petal length	Petal width	Sepal length	Sepal width	Petal length	Petal width	Sepal length	Sepal width	Petal length	Petal width
5·1	3·5	1·4	0·2	7·0	3·2	4·7	1·4	6·3	3·3	6·0	2·5
4·9	3·0	1·4	0·2	6·4	3·2	4·5	1·5	5·8	2·7	5·1	1·9
4·7	3·2	1·3	0·2	6·9	3·1	4·9	1·5	7·1	3·0	5·9	2·1
4·6	3·1	1·5	0·2	5·5	2·3	4·0	1·3	6·3	2·9	5·6	1·8
5·0	3·6	1·4	0·2	6·5	2·8	4·6	1·5	6·5	3·0	5·8	2·2
5·4	3·9	1·7	0·4	5·7	2·8	4·5	1·3	7·6	3·0	6·6	2·1
4·6	3·4	1·4	0·3	6·3	3·3	4·7	1·6	4·9	2·5	4·5	1·7
5·0	3·4	1·5	0·2	4·9	2·4	3·3	1·0	7·3	2·9	6·3	1·8
4·4	2·9	1·4	0·2	6·6	2·9	4·6	1·3	6·7	2·5	5·8	1·8
4·9	3·1	1·5	0·1	5·2	2·7	3·9	1·4	7·2	3·6	6·1	2·5
5·4	3·7	1·5	0·2	5·0	2·0	3·5	1·0	6·5	3·2	5·1	2·0
4·8	3·4	1·6	0·2	5·9	3·0	4·2	1·5	6·4	2·7	5·3	1·9
4·8	3·0	1·4	0·1	6·0	2·2	4·0	1·0	6·8	3·0	5·5	2·1
4·3	3·0	1·1	0·1	6·1	2·9	4·7	1·4	5·7	2·5	5·0	2·0
5·8	4·0	1·2	0·2	5·6	2·9	3·6	1·3	5·8	2·8	5·1	2·4
5·7	4·4	1·5	0·4	6·7	3·1	4·4	1·4	6·4	3·2	5·3	2·3
5·4	3·9	1·3	0·4	5·6	3·0	4·5	1·5	6·5	3·0	5·5	1·8
5·1	3·5	1·4	0·3	5·8	2·7	4·1	1·0	7·7	3·8	6·7	2·2
5·7	3·8	1·7	0·3	6·2	2·2	4·5	1·5	7·7	2·6	6·9	2·3
5·1	3·8	1·5	0·3	5·6	2·5	3·9	1·1	6·0	2·2	5·0	1·5
5·4	3·4	1·7	0·2	5·9	3·2	4·8	1·8	6·9	3·2	5·7	2·3
5·1	3·7	1·5	0·4	6·1	2·8	4·0	1·3	5·6	2·8	4·9	2·0
4·6	3·6	1·0	0·2	6·3	2·5	4·9	1·5	7·7	2·8	6·7	2·0
5·1	3·3	1·7	0·5	6·1	2·8	4·7	1·2	6·3	2·7	4·9	1·8
4·8	3·4	1·9	0·2	6·4	2·9	4·3	1·3	6·7	3·3	5·7	2·1
5·0	3·0	1·6	0·2	6·6	3·0	4·4	1·4	7·2	3·2	6·0	1·8
5·0	3·4	1·6	0·4	6·8	2·8	4·8	1·4	6·2	2·8	4·8	1·8
5·2	3·5	1·5	0·2	6·7	3·0	5·0	1·7	6·1	3·0	4·9	1·8
5·2	3·4	1·4	0·2	6·0	2·9	4·5	1·5	6·4	2·8	5·6	2·1
4·7	3·2	1·6	0·2	5·7	2·6	3·5	1·0	7·2	3·0	5·8	1·6
4·8	3·1	1·6	0·2	5·5	2·4	3·8	1·1	7·4	2·8	6·1	1·9
5·4	3·4	1·5	0·4	5·5	2·4	3·7	1·0	7·9	3·8	6·4	2·0
5·2	4·1	1·5	0·1	5·8	2·7	3·9	1·2	6·4	2·8	5·6	2·2
5·5	4·2	1·4	0·2	6·0	2·7	5·1	1·6	6·3	2·8	5·1	1·5
4·9	3·1	1·5	0·2	5·4	3·0	4·5	1·5	6·1	2·6	5·6	1·4
5·0	3·2	1·2	0·2	6·0	3·4	4·5	1·6	7·7	3·0	6·1	2·3
5·5	3·5	1·3	0·2	6·7	3·1	4·7	1·5	6·3	3·4	5·6	2·4
4·9	3·6	1·4	0·1	6·3	2·3	4·4	1·3	6·4	3·1	5·5	1·8
4·4	3·0	1·3	0·2	5·6	3·0	4·1	1·3	6·0	3·0	4·8	1·8
5·1	3·4	1·5	0·2	5·5	2·5	4·0	1·3	6·9	3·1	5·4	2·1
5·0	3·5	1·3	0·3	5·5	2·6	4·4	1·2	6·7	3·1	5·6	2·4
4·5	2·3	1·3	0·3	6·1	3·0	4·6	1·4	6·9	3·1	5·1	2·3
4·4	3·2	1·3	0·2	5·8	2·6	4·0	1·2	5·8	2·7	5·1	1·9
5·0	3·5	1·6	0·6	5·0	2·3	3·3	1·0	6·8	3·2	5·9	2·3
5·1	3·8	1·9	0·4	5·6	2·7	4·2	1·3	6·7	3·3	5·7	2·5
4·8	3·0	1·4	0·3	5·7	3·0	4·2	1·2	6·7	3·0	5·2	2·3
5·1	3·8	1·6	0·2	5·7	2·9	4·2	1·3	6·3	2·5	5·0	1·9
4·6	3·2	1·4	0·2	6·2	2·9	4·3	1·3	6·5	3·0	5·2	2·0
5·3	3·7	1·5	0·2	5·1	2·5	3·0	1·1	6·2	3·4	5·4	2·3
5·0	3·3	1·4	0·2	5·7	2·8	4·1	1·3	5·9	3·0	5·1	1·8

2.1 Table 1 from R. A. Fisher's 1936 paper "The Use of Multiple Measurements in Taxonomic Problems," organizing, into rows and columns and as a complete dataset, the measurements of petal and sepal width and length of 150 different irises from three separate species that grew on the Gaspé Peninsula in Quebec, Canada.

> ***Pop-Up* Definition: Data Structure**
>
> A ***data structure*** is a term used by computer science to designate any stabilized form that gathers and organizes data objects. These objects can include data values, the relations among these values, and the results of organizing data and their relations. A data table is a kind of data structure in which a collection of values or data points is organized in rows and columns. Each column specifies a particular attribute or dimension of the data. Each row is a set of data points corresponding to an instance being measured for each of the attributes. The table is not only a storage space for data points but a way of arranging attributes and the ways they are being measured. A data structure such as a table, therefore, also implies a way either for the data to potentially *operate* internally to the structure or for it to become calculable.
>
> **Data structures** are often distinguished from algorithms; the latter are defined as modes of acting on data structures. A table collects, orders, and stores the data, but an algorithm can then be used to sort the same data into ascending or descending orders. However, data structures and algorithms are co-compositional. The data table that structures the iris dataset contains two levels of column: a species level (upper), and a sepal and petal measurement level (lower). By aligning the measurements or data points into rows and columns, the data table visually organizes the data for comparison and contrast, displaying the relations between the upper-level species' columns.

y coordinates for each specific point or measurement. Once the *Iris* data are tabulated, this vectoriality becomes obvious: the first row of the *Iris setosa* is a numerical vector of 5.1, 3.5, 1.4, 0.2, which organizes the data in terms of sepal width and length to petal width and length comparisons. These are not isolated measurements, then, but positions held among individual specimens and then across species through a series of relations. The table prefigures the inputting of computational data for ML; a data "object" such as digitized image or a sequence of letters being inputted into a neural network is also already a numeric vector (and/or series of these vectors), even if it is also just a single data point, since the entire operation of ML depends on data points becoming relatable across a neural network.

Data tables are one kind of format that commonly features in statistical ways of seeing the world. They are also a computationally typical mode of formatting data points so that they can be called on by algorithms and models. They are the basic mode of storing and organizing data points from graphic interfacial programs such as Excel to relational databases such as MySQL. I want to hold on to this primary relationality that makes data the computable "stuff" of which ML is made operative. Data cannot be prized apart from the way they are made operable for an algorithm, model, or neural network. And this making operable occurs not simply through cleaning and optimizing data but, importantly, through formatting it via modes of (visual) arrangement and display. However, the organization of data into mobile patterns of relation across and up and down a data table, for example, reveals that questions of display are not the only issues and movements reworking the "bare" measurement of data.

Fisher's table seems to stand up as a self-contained, desaturated, and neutral approach to enumerating the world. But its seemingly uncolorful tabulation of *Iris* data, and subsequent availability and dissemination as a "universal" and ubiquitous dataset, means that it belongs to a culture of computation in which questions of data bias and debiasing are increasingly posed. The claim, especially proffered by the tech sector, that "poor" data are the underlying problem biasing an algorithm or model toward racism or sexism too easily skirts over the primacy of relation in ML. The implication here is that while the data are the quanta on which an algorithm performs its "learning," bias lies with the data alone. Racial bias in a facial detection algorithm that cannot recognize Black or Brown faces is due to an overrepresentation of white faces in the model's training dataset, for example. To be fair, the work done by researchers such as Joy Buolamwini and Timnit Gebru (2018) has pointed to how the overrepresentation of white and male bodies, faces, and skin in datasets used in computer vision leads to a pipeline of biased embeddings throughout the algorithmic chain. This can then develop into serious errors in, for example, the automated detection of melanomas in clinical settings where models have not been trained on data with dark skin samples (2018, 2). The importance of such work on racial bias in datasets should indeed be acknowledged and lauded. But at the same time, this approach to bias leans heavily toward data representation and labeling solutions to debias the racism of the model. This leaves unanswered questions about how an *agencement* both assembles out of and generates new racialized knowledge. As we shall see, Fisher's *structuration and operationalization* of data and algorithm are crucial to the color lines that run through statistics. The statistical genealogy of ML normalizes the whiteness of its color line and actualizes its axiomatic functions as already racialized.

Data, Whiteness, and Blackness

It has taken platforms and the ML community quite some time to acknowledge the problem of algorithmic racism (or sexism, or . . .), especially the extent to which this goes beyond the data as such. The year 2018 proved to be something of a watershed for platforms as they began to acknowledge a bias problem in their models' technicity. Subsequently, a flood of debiasing solutions was released and publicly aired. Facebook announced "Fairness Flow" at its developers' conference in May 2018. Fairness Flow was an automated system for measuring how Facebook's own AIs interact with specific groups of people and was to be inserted into the backbone of the platform's software-engineering workflow (Kloumann 2018). In July of the same year, Microsoft Research published a blog post describing a fairness algorithm for binary classification systems used in, for example, deciding whether A or B groups of people might be a better set of candidates from which to select for hiring (Dudik 2018). This post was especially telling in that its accompanying banner image suggested that Microsoft was aware which racial groupings might already be "outliers" of classifier algorithms. Google followed suit in September 2018, releasing its What-If developer tool kit, which can be used to analyze the input of its deep learning model TensorFlow and visually represent how decisions come to be made algorithmically (Wexler 2018). Soon after, IBM announced that it had one-upped Google by releasing visual developer tools that were platform independent in its AI Fairness 360 tool kit (Linux Foundation 2020), enabling developers to analyze any ML model for bias. All these solutions to bias increase the operations of ML on . . . ML. In many ways, this intensive becoming-operational of ML—the production of machine readability and learnability for machines by other machines—is suggestive only of a typical engineering approach to sociotechnical problems. On the other hand, it also indicates just how pervasive the autopoietic tendencies of the ML assemblage are. However, the fact that bias is not so easily fixed either by improving the data or by implementing technical fixes across the data pipeline is borne out by how "color" is quite literally dealt with by ML's deepaesthetics.

The much-commented-on instance of Google Photos' (an app that uses deep learning) labeling a photo of Jacky Alcine's Black friends as gorillas in 2015 as he performed a search on his phone is just such an operation of data-algorithm relations enacting a racism particular to ML experience.[4] Alcine, a software programmer, called out this labeling as racist but also pointed the finger at Google's sample data on which the app trained. There is no doubt

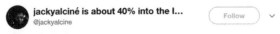

2.2 Jacky Alcine's tweet from 2015, showing the way Google Photos had grouped a picture of him with a friend under the label "Gorillas."

that training data are indeed too homogeneous in their nonrepresentative sampling of, for example, images of Black, Brown, and Asian people, but we also need to understand that the algorithmic production of "labels," or classes, become limit cases for any new data inputs into the model. Data are thus not a pool to be diversified, dissociable from the operations of the model. Rather, data are always already in the process of becoming machine learnable at various points of the development of an AI's technical ensemble.

In the United States and United Kingdom, the overwhelming understanding that has emerged among Black users of social media is of a visual culture that is generatively entangled with and produced by the overt *and* unseen (by white seeing) image operations of platforms such as Google and Facebook. Commentary or affective reactions found in tweets such as Kabir Alli's videoing with his friends and their collective guffawing at his search results returned by Google for the phrase "three Black teenagers," or Alcine's palpable anger at the "gorilla" labeling, suggest familiarity with the ways racism is made operative by platforms.[5] The frequent return of police mug shots of Black people in online searches for "Black man" or "Black youth" connects the visual cultures of platforms and the lived experience of

> ***Pop-Up* Definition: Labels and Classes**
>
> Both **labels and classes** are used to enable categorization of data, and the terms are sometimes used interchangeably in data science. Labels are sometimes applied manually by humans to categorize samples from a dataset so as to provide training data for a model. For example, 10,000 images might be individually labeled to sort them into different dog species. The labeled images are then used to train a model to recognize and pair a large dataset's images with these labels. This then allows the model to accurately output unlabeled images into the class "red setter" or "labrador." This human-machine method of manually labeling and then computationally training was initially used to structure and produce the ImageNet dataset, for example.
>
> As Jacky Alciné's callout of the class "Gorilla" to which his friend had automatically been assigned or labeled by the Google Photos app demonstrates, the pipeline from data input to output is both highly optical and invisual—the making operational of visual aspects of data—in ML. Automated and trained relations between the weights and biases of neural networks mean that data and algorithmic legacies from datasets, data structures, and prior functions all ingress into processes of categorization taking place in a wide range of ML-based image recognition apps and AI.

photographic genres of criminality, which circulate between police databases and news and media outlets. Data practices of classification and labeling are also historically well-known and understood through lived relations to such photography, museal forms of display (that feature phrenology, for example), and the widespread image traditions of the British and American eugenics movements. Refinding such practices in the operations of artificial intelligence software, then, resonates with the visual skills and knowledge many Black and people of color will already have acquired. In some ways, it would be easy to assign the racism of online platforms a simple continuity with an already circulating white visuality, ramped up on the steroids provided by neural network architectures.

Yet the operational difficulties in "diversifying" a normalized whiteness become starkly obvious when platforms themselves attempt to counter or resolve allegations of racism. When Google acted in response to Alciné's call on its racist labeling, the company's solution, which took three years to bear fruit, was *not* to add more diverse images of Black people to its own dataset—a solution that, arguably, might have furnished a wider sampling of skin tones and generate more variability of color. Instead, Google removed the label "gorillas" from the image classes that Google Photos' model deployed to recognize photos (Simonite 2018). This suggests that Google Photos was unable to sufficiently adjust differences in its feature detection for the features being learned and recognized for that label. It was unable to maximize variation *within* a label/class while maintaining distinctions between labels/classes. In other words, the variation within a label such as "friends" was perhaps not heterogeneous enough in terms of its color features to match across the range of differing images of the blackness of people within Alcine's photo archive. But at the same time, the shape of his friend's face—registering potentially through a feature such as edge detection—was too variable for the model's "friend" label, potentially trained on the nonvaried data of mainly white face shapes. This in turn makes it a feature of a different class, such as "gorillas." My determinations here are speculative, given that neither we nor perhaps even Google's engineers know what is happening at the level of feature detection in a complex and many-layered deep learning model. But I offer such speculations because they indicate that adding difference to the data alone is insufficient, since the data are also delimited. Such delimitation enables the production of a class or label bound by the relations between the features the neural net has extracted and its subsequent use of those features to further discriminate or classify future data. Algorithm and data autopoietically iterate. As Alcine notes, the question of the color black, while too variable for the Google Photos label, is historically too "pure" in its degree of blackness to be classed as human: "'It says "gorilla," and I'm like "nah,"' Jacky says. '[It's] a term that's been used historically to describe black people in general. Like, "Oh, you look like an ape," or "you've been classified as a creature,"' Jacky said. 'Because the closer they looked to a chimp . . . the more black, the more pure the blackness was supposed to be'" (Alcine, quoted in Poyant et al. 2015). Alcine's comments reinforce the ways in which the representation of Black bodies has been rendered historically via a white subject/face/body maximizing its purchase on the entirety of humanity. But something else is also coursing through color; while always in relation to bodies, histories, and technics, color is *crucially* relational and always emerging and experienced with other colors. As Erin Manning (2008, 330) notes: "We never see a color as such. We experience a worlding of color that resonates between different coloring tendencies." Machine learning also never sees color as such because it is constantly transcoding color into arrays of numerical values whose matrices are distributions of the frequency of some pixel values compared with others. And yet the worlding it has been involved in producing does not preserve the generative relationality of coloring to which Manning points. Instead, it funnels this relationality through the operations of typology. Artificial intelligence models, such as the ones powering Google's photo app, render color and other features according to typological schemas—classes, labels, and so on—made possible through statistical operations such as frequency distribution. As Ramon Amaro (2019) suggests, "Such a typology re-slices through the tendencies of color as messier, 'worlding' relational fields."

Understanding the ways in which, as I have suggested in the previous chapter, ML as *agencement* is multiplicatory and conjunctive will help us to better conceive how the quantitative data with which ML deals are not merely numerical aggregates. They are already in processes of assembling via immanent *quasi*-qualitative relations. I want to suggest that some of the qualitative aspects—which at their primary and most imperceptible register consist of relations of proximity and distance between and among different kinds of statistical vectors—become the register through which ML actualizes a racist imaginary across platform cultures. As such, it becomes impossible to simply replace racist data or add in more "color" to datasets to correct racism. Racism, I will argue, lurks in the numerical, spatial, as well as visual conjunctions of ML, in which a statistical imaginary, whose genealogy is seeded in its proximity with eugenics, functions as the associated milieu for its technical ensemble.

At the same time, the very relationality of data and operations core to ML's functioning also suggests that the question of "color" and its racist actualizations are at least contingent and can be reengaged as mutable, allowing them to "world" differently. How an indeterminant, emergent, and differential blackness for ML's statistical imaginary might be otherwise realized opens in the work of artists such as Stephanie Dinkins. Using a deep learning network on a small dataset comprising oral histories of three generations of Black women (from her own family), work such as *Not the Only One* (2018–) encourages the AI to become relationally across data provenance, scale, and technical "quirkiness" (Dinkins 2018). In Dinkins's practice, we see a refashioning of ML as a different kind of assemblage, whose conjunctions hold through relations of experiencing and experiences of making Black life through her family, but also through broader participation with Black and people of color AI communities. This seeds the possibility of a blackness for ML that is no longer determined solely by a statistical and typological "coloring" (see fig. 2.3). Instead, AI would lean in toward a blackness that

> How to make an AI robot from scratch*....
> Getting started:
> 	learn Tensorflow
> 	test deep writing neural network using Toni Morrison's Sula as data
> 	interview source subjects (create data)
> 	test deep writing neural network using Toni Morrison's first interviews
> 	test neural networks (algorithm) options
> 	make algorithmic output make sense
> 	record more interviews
> 	record more interviews
> 	develop more incisive questions
> 	record more questions
> 	recruit POC programmers, technologies to join team
> 	master Tensorflow

2.3   Screenshot of Stephanie Dinkins's procedure for "How to make an AI robot from scratch." Artist's website.

is primary, constitutive, and productive of a pluralistic universe. This, Fred Moten affirms, is a "nonexcluded, nonexclusive understanding of mixture, of color, as constitutive of blackness and of blackness or black as a constitutive social, political, and aesthetic power" (2008, 193).

Operationalizing Race through the Statistical

Something about the stark facticity of the measurements of Fisher's *Iris* dataset requires further attention before proposing a novel decoloring or recoloring for statistical "vision" such as Dinkins offers. In Fisher's Table 1 (see fig 2.1), the colors and relations are precisely a black-and-white (color) line. As we shall explore soon, ML's racism has a genealogy enmeshed with the nineteenth-century genesis of statistical techniques proper, in which Fisher's work played a key role. The intimate sociopolitical relations between statistics and eugenics have been noted by historians of statistics such as Alain Desrosières (1998, 259–63), who presents their relation within a broader framework of understanding a history of science via transductions across epistemologies, practices, and technical apparatuses. He argues that the eugenicist statistician Karl Pearson enfolded a theory of correlation into the political program of selective racial and class-based breeding (that is, eugenics) via a mathematical system of statistical techniques. Desrosières demonstrates how statistics not simply was accidentally in relation to a program for social engineering but was from its outset a way of *knowing* and *acting* on the world to engineer it. I will draw on Desrosières's insights into how the shift to correlation and away from mechanical causality in the early eugenicist statisticians initiates a privileging of pattern and association between data and empirical phenomena. This continues to permeate ML, especially in image recognition. On the one hand, then, contemporary ML-based AI operates in starkly racist ways along a color line, in which image recognition tasks and processes have literally failed to detect Black bodies and faces or Asian facial features. On the other hand, the racism of statistics passes by unnoticed in so many ML contexts because it is not overtly *causally* connected to the production of racist images or even race-based categorizations per se. At the core of ML's racism is a statistical genealogy of typological thinking-making whose immanent spatial relations—its mode of positioning phenomena as proximate or distant from each other—strive for the reduction of variability within groups, categories, classes, and ranges. At the same time, the difference/distance *between* classes and ranges increases—is maximized—often via vectorial techniques.

The development of techniques for dealing with the reduction of data variability *within* to maximize difference *across* its typology begins via the entanglement of statistics and eugenics. However, the genealogy that I want to trace from eugenicist statistics to ML is less aligned with arguments attributing their imbrication to an ascendant biopower or their inmixing that has haunted twentieth-century genetics.[6] The problem with this biopolitical emphasis is that it accords too much weight to the visibility of biological bodily traits as key to the perpetuation of racism. This tends to let statistics as a racist *agencement* off the hook; eugenics' hold on statistics "fades away" historically in many accounts of its overtly racist program of selective breeding (Louçã 2009; Desrosières 1998, 146). As Aubrey Clayton has argued, eugenics and statistics are deeply imbricated at the level of an ongoing visible set of characters, figures, and institutions—Francis Galton, Karl Pearson, and Ronald Fisher, and the laboratories and universities they established and which have continued to uphold their reputations—and at what she calls "less visible ones [that] are embedded in the language, logic, and philosophy of statistics itself" (Clayton 2020). I want to propose that it is not *bio*politics that takes hold of bodies and produces a racism via its coincidental historical intersection with statistical techniques. Instead, statistics is a program *for* calculating bodies (and other phenomena) as entities capable of being qualitatively redistributed according to quantitative typologies. Typology makes quanta operative for statistics; numerical operations come to produce race through various functions that follow and generate vectors of quanta-qualia, amid spectra of categorical determinations, which also edge on to class, gender, sexuality, and myriad subjectivations. The functioning of such vectors is often imperceptible, although they will have visible corporeal ramifications. But if we only see racism operating in statistics at the level of the bios, then we miss its invisual architecting—its making of "race" *in* the ML relations that condition contemporary computational visuality.

Wendy Chun's conception of race as technology or technique can be considered in this context, since she emphasizes not what race is in terms of bodies and their given determination by race or even by ra*cism* but rather *how* technologies and techniques *race*. She takes eugenics as a test case, in which its acceptance of Mendelian genetics became a technique for both separating and suturing culture/nurture and biology/nature. For her, eugenics both traversed programmable technical solutions for sociality via statistical means *and* posited that genes lay outside the realm of human control (Chun 2011, 120–24). Yet the becoming-statistical of eugenics also makes its racism difficult to pin down, since, as contemporary commentators point out, eugenics was never about privileging whiteness per se (see, e.g., Cain 2019). Paradoxically, it is statistics' seemingly neutral interest in quanta that cloaks the lines of its coloring, which are nonetheless definitively concerned with the maximization of purity. Chun's interests, at any rate, lie with how race becomes a full-blown technical program under a broader regime of biopolitics. Mine are concerned with how ML operationalizes racism in singular ways.

Ruha Benjamin gives us a sense of how this occurs through what she calls AI's "architecture" of racism in which its sorting, classifying, and predictive techniques afford it durability for white sociality (2019, 166). Like Chun, Benjamin understands technologies such as ML as generative or productive of race and racism rather than tools that simply serve a racist agenda. Predictions in Black policing and criminal DNA profiling and matching, she argues, always adhere to a "group" (class or label) whose frequency distributions are broad enough to become generic. The generic image of a potential Black criminal predicted by an AI running matches and convolutions across a police database—whose data are extracted from an archive of those already policed racially—also creates room for a range of many new potential matches because of its very genericism. In this way, a "light-skinned Black man" is both targeted *and* generic, for example (see also Sankar 2010, 57). I want to take off from Benjamin's analysis of race after computation, especially machine learning, as an architecture but heed its slippery mobility and its ability to move information rather than adhere to plans or blueprints. Such architect*ing* relies on an arsenal of statistical techniques such as frequency, distribution, and dimensionality reduction to glue it together. When these techniques conjoin with the convolutions, feed-forward, and back-propagating movements of neural networks, an amorphous assemblage gets underway. Such an architecting does not have to actively draw on databases that misrepresent or underrepresent Black people or people of color, nor does it have to strictly enforce or perpetuate a color line, although it may indeed do all these things. What it strives for instead is a programmatics that tries to maximize an evenly distributed range of invariance for its categories/labels/classes. Such classes must be able to span some variability but, at the same time, provide parameters *and* a constant range of distribution for what varies in their membership. Eugenics has played a crucial role in architecting this program because it materialized race as a problem to be addressed in terms of how to both account for and constrain the internal variability of classes or groups of (genetic) traits, toward the goal of maximizing and optimizing each group's *invariance* in relation to another group. While new deep learning–trained AIs do not necessarily deploy classes, nonetheless the legacy of this *discriminant analysis*—either because of the direct deployment of adjunct discriminant algorithms in the training or optimization of a model or because of the typology of datasets that models such as CNNs were originally trained on—sets up the potential to *racialize* the model. The only thing that can outrun the model is color itself, since its variability—that is, its refusal to be reduced to a quantitative measure, since colors are tendencies of experience in race, perception, and technics—will always insist on indeterminacy.

It may, then, be more useful to understand race, generated at/in the intersection of eugenic and statistical practices, as less than a fully elaborated "architecture" or technology, since this paints it as more static and hence more easily circumscribable. The statistics-eugenics nexus processually generates race instead via honed practices and *techniques*. An emphasis on technique moves us toward an analysis of race and technology that does not aim to reveal a generalizable evaluation; that is, ML or statistics are identifiably racist. This is not to say this *isn't* the case, and there are plenty of instances in which algorithms and AIs result in racist determinations. Alongside Amaro's (2023) call to think of the algorithmic as a mode of existence in which the

"Black technical object" might individuate differently (rather than simply be racially erased), we can work in different directions simultaneously. We can acknowledge the ways in which race becomes substantialized and open to quantification, but also seek where color, in its infinite variability, might also be "cultivated" (Simondon 2015, 17) via *techniques*. Here race and color are made both operatives in the formation of racism and powerful social, political, and aesthetic forces for *singular*, novel change. Every technique is both singular and general, as Manning suggests: "There are techniques for hoeing, for standing at a bus stop, for reading a philosophical text, for taking a seat in a restaurant, for being in line at a grocery store" (2012, 33). There are particularities to standing in line: you stand up straight as you approach the head of the queue at immigration control despite the jet lag setting in. The queue, then, is both a generalized composition and a formation for attending to each member, one at a time. Specificity and generality imbricate each other in technique. To generate techniques is to already be in the middle of a complex ecology of practices that takes us outside of ourselves, placing "us" in an already composing relationality. Understanding how statistical and ML techniques produce racism also allows us to continue to prehend ML experience relationally.

Figuring Statistical Racism

I want to approach the contemporary problem of racism in ML genealogically, closely eyeing two of its "figures" or functions: principal component analysis (PCA) and linear discriminant analysis (LDA). Both are less flashy operations than fully formed AI models such as StyleGAN or DeepFace; they function more as workhorses of the ML arsenal. As it turns out—and not coincidentally, as we shall come to see—PCA was a function first proposed in 1901 by the statistician and eugenicist Karl Pearson. On the other hand, LDA returns us to Fisher, finding its epistemic kernel in the problems of discriminating between classes in the multiple measurements of the *Iris* dataset.[7] Fisher was likewise a eugenicist. Both PCA and LDA are routinely deployed in ML for dealing with problems of data quantity and distribution, and their use is so mundane that they are barely remarked on. Yet they offer an axis of tendencies that set out major directions for how machine learning–driven AI has, does, and will unfold. Principal component analysis follows a vector in which variability is understood through comparing maximal and minimal distances in similarity between components of the data. Linear discriminant analysis, on the other hand, is concerned with the creation of classes that tolerate internal variance in membership but are nonetheless maximally differentiated from other external classes. In just these two functions, then, we already have techniques, which are plastic and typological, making them both the harbingers of and ongoing operators for contemporary machine learning assemblages.

As I hope to show, reengaging these statistical functions in relation to both ML operations and eugenic genealogies can open ways of transversally engaging with and intervening into ML as *agencement*, that is, as a machinic multiplier of not only technical but also race relations. To pursue such a genealogy of ML via a statistic-eugenic nexus is to understand how its racism cannot simply be washed, debiased, or data cleaned even. ML continues to individuate through a sociotechnical project and mode of querying (empirical) experience that conditions it to quantify variability, and specifically racial variation, through both typological and vectorial means. As Amaro notes more broadly: "The conditions for racial sorting and priority were already set forth in the establishment of data analytics and statistical correlation as viable tools for social inquiry. In many ways, the limits of calculation cannot be understood outside of the connection between racial sorting, social welfare, and quantification" (2019). As we will see, this fundamentally shifts the ways in which race comes to be "seen" through the distribution and constraint of parameters of variability and as computation is reconfigured via these statistical operations.

In the opening line of Pearson's paper on the algebraic formulation of PCA, he states, "In many physical, statistical, and biological investigations it is desirable to represent a system of points in plane, three or higher-dimensional space by the 'best-fitting' straight line or plane" (Pearson 1901, 559). The PCA function performs a transformation of variables that have first been represented as a system of points—a data table such as Fisher's *Iris* data could form one such system—on an x, y coordinate system. These are then shifted to another kind of space, no longer matrixal or grid-like but instead *vectorial*. The result, algebraically, is to extract ordered "directions of change" or "principal components" across all the data being analyzed.[8] These directions indicate the overall maximum to minimum directions of variance of the data *as a set*. The importance of PCA as a germinal technique for ML lies in the way it organizes features that are extracted from the dataset. It sets up the potential for "principal" features to be arranged in relation to each other through directions of maxima and minima proximity/similarity and distance/distinction without using a pregiven set of classes or labels to classify the features. Vectors and relations come to contour the data, gener-

2.4 The use of PCA in facial recognition. The top row depicts the original inputs to the PCA algorithm; the bottom row depicts the eigenfaces, or vectorial images, that represent ranges of maximum variance across the entire original dataset.

Pop-Up Definition: Principal Component Analysis (PCA)

Principal component analysis (PCA) is a common dimensionality reduction algorithm. Data that are inputted to the algorithm have more than one dimension (usually high-dimensional data such as images with many attributes) and output a dataset of smaller dimensions. This smaller dataset is made up of the principal components of the entire original dataset in the form of a structured set of vectors. The vectors are descending ranges of the maximum variance of the original data, ordered from most to least variation. This vectoral set can be used inversely to approximately reproduce the original data inputs.

Using PCA in facial recognition, a larger training dataset of m number of images is reduced to a new set of representational images that constitute the principal image directions across all the images in the training set. These are known as *eigenfaces* and are stored as vectorial images. They provide the reference set to which new (future) image inputs are then referred for recognition. Principal component analysis is an "unsupervised" ML algorithm, which means that none of the inputs or outputs use labels or predetermined categories.

ating it as a particular kind of "set." As we have already seen in the previous chapter, the same logic of feature extraction and vectorial relationality runs through more complex neural networks such as CNNs.

From the 1980s onward, PCA was frequently used as a dimension reduction algorithm, especially as datasets substantially increased in size and variability. And from the late 1980s to the mid-2000s, PCA was the primary function deployed in ML facial recognition. Although PCA has been superseded by the multilayered deep learning approach of models such as Facebook's DeepFace, it is nonetheless routinely used in conjunction with a raft of neural networks as a preceding operation on, for example, training data.[9] It is also incorporated into larger AI assemblages to organize data points at initial layers of a network. The computational success of PCA lies in its capacity to use a smaller dataset representative of the initial higher-dimensional one that maintains most of the key information about the dataset's variance. This representational capacity manifested visually in PCA's use in early facial recognition models. This makes its use during this period interesting to revisit, given that neural architectures processing data are now literally too massive and distributed to be humanly viewable. In earlier ML facial recognition, then, PCA almost allows us to "see" the visuality of a stark black-and-white procedure of ML feature discrimination. In this context, PCA provides a snapshot of how discriminant analysis worked before ML's subsumption into the nonoptical vectorization of space accomplished by deep neural models.

In these earlier approaches to facial recognition (e.g., Turk and Pentland 1991; Belhumeur, Hespanha, and Kriegman 1997; Calder et al. 2001), PCA vectorially extracted the main variations—the principal components—across an initial training dataset of face images. This then produced a second, significantly smaller set of image data comprising only these main "feature directions" (fig. 2.4).[10] Hence the distributions of variation across the data as a whole are maintained while significantly reducing the quantity of data. This new image dataset was then known as "eigenfaces," after the *eigenvector analysis* to which it had been subjected. It is this new set of eigenfaces that are then said to make up a "face space" (Turk and Pentland 1991, 74), against which new and future image inputs are compared and then "recognized." The original training dataset images can also be reconstructed, with varying success, sui generis by drawing on and combining weighted proportions from each of the eigenfaces. In these operations, images are being added to, multiplied, and recombined in specific ways through nonhuman processes of observation such as distribution determination, weighting, and recombination. The first eigen image extracted via PCA from a facial dataset will show the "component" or vector that is the most dominant feature direction of the whole image dataset. A typical component might simply be the amount of brightness or light cast over part of an image. The total collection of eigen images contains information about the dataset's primary vectors: "The projection operation characterizes an individual face by a weighted sum of the eigenface features, and so to recognize a particular face it is necessary only to compare these weights to those of known individuals" (Turk and Pentland 1991, 71). Used in this way, PCA generates a new kind of image, made up purely of information about other images. An eigenface does not *picture* an empirically existent face but is a visual rendition of vectorial metarelations. The set of eigenfaces operationalizes the activity of looking at and extracting knowledge from some phenomenon as well as becoming a substrate for ongoing observation. While a reduction of dimensionality is being performed, an aggregation and modulation of the variability of images in the original dataset also take place. Eigenfaces constitute a new kind of informatic aggregate that stores observations about another set of images, creates a directional range of values for selection/observation operations on new (incoming) image sets, and redistributes observational operations so that discrete selections are created as outcomes of aggregate operations such as averaging and sweeping of a hierarchy of diminishing importance of components. We should recall here the logic of racial profiling, raised by both Benjamin and Sankar, entailing both a generic casting of its net and a specificity of targeting. The same logic is at work in the aggregate and selective operativity of PCA.

Vectorizing Race

To an extent, PCA has become less relevant for image recognition since the advent and relative success of large-scale image and language models in computer vision tasks. However, deep learning architectures typically borrow and combine functions such as PCA or other forms of vector analysis such as linear discriminant analysis. Yet from the viewpoint of the relation that I am suggesting is crucial between image aggregate and the setup of vectorial conditions necessary for training a model, CNNs, for example, can also be considered as more sophisticated developments of the same architectural logic of functions such as PCA. They continue the transformation of visual-indexical-extensive space figured by x, y coordinate systems to nonvisual-vectorial calculability, which I drew attention to in chapter 1. A CNN is an expanded form of operations in which automated distributed observation performs on and modulates datafied images.

Principal component analysis in facial recognition literally visualized a "subspace"—or face space—of ML. We have already encountered such subspaces as latent, in chapter 1. As Matteo Pasquinelli suggests with respect to neural networks, subspaces "calculate a statistico-topological construct" (2017). In its deployment of statistical induction to create functions in relation with inputs, ML transforms data from any empirical hold on phenomena to a different calculative "entity" comprising relations, forces, and degrees of a topology. The calculative "subspaces" and statistico-topologies of ML share continuity with a certain sociopolitical diagram, which, borrowing from Deleuze and Guattari, we can term "facialization." Functioning according to two mechanisms, the "machine" of facialization first produces a reference surface based on mechanisms of inclusion and exclusion: "*Regardless of the content* one gives it, the machine constitutes a facial unit, an elementary face in biunivocal relation with another: it is a man *or* a woman, a rich person or a poor one, an adult or a child, a leader or a subject, an x *or* a y" (2005, 177). We might, for example, locate in the vectorial production of "face space" by PCA, or in an early layer in a CNN for filtering features, the operative components of just such a surface of reference. The second aspect of facialization works to compare new images to this space, or as Deleuze and Guattari argue: "Under the second aspect, the abstract machine of faciality assumes a role of selective response, or choice: given a concrete face, the machine judges whether it passes or not, whether it goes or not, on the basis of the elementary facial units" (177). Face space in PCA—and other kinds of subspaces that permeate ML—are just such surfaces of reference operating under a reconfigured *automated* regime of facialization. Likewise, new "faces," and indeed any new image instances, are entrained via a facialized regime that does not index the visible qualities and characteristics of actual faces but relies on an invisible smooth vectorial surface, a hidden layer(s) immanent to the model's own computational architecture. Crucially, facialization is a *coloring machine*: black holes on a white face puncture its surface, producing the reference or color line and indeed entire surface for what is allowed to pass and what is not.

The logic of PCA partakes in the eugenicist project less in terms of its contribution to biometric governance and more as a topological operation that enacts functions for selection. It operates to produce a range of variance for what can and cannot pass social criteria such as "good breeding" and "superior stock." Pearson actively worked toward implementing a program of socially engineered "practical eugenics" (1912), which he saw as the bridge between his work in statistics and his desire for racially selective population management. Statistical techniques would provide evidence of correlative trends supporting the truth that survival of the fittest could only be accomplished by a regulative program of good breeding (1912). It is also worth noting that Pearson's statistical work was responsible for the calculation of nominal variables—for example, how to sort, classify, and organize discontinuous data such as variations of eye color in children who should nonetheless belong to the same "race." He argued that visible variations were subtended by an unobservable "normal distribution curve" across the entire population (see Kevles 1995, 31). Like the vectorial features of PCA, the curve did not simply demonstrate a distribution of difference but determined the range of variations across a population. Hence a program for social selectivity comes to be rendered via invisual topoi such as this distribution curve—a statistical pattern that operates to render, subordinate, and manifest a range of "appropriate" variation. In this unobservable distribution curve, we have the germ of the sub- or latent space. In a time of machine learning–driven AI, these spaces become full-fledged surfaces of dynamic operativity and ones for sociotechnical determination.

Eugenics, of course, not only was racist but extended its call for demographic selectivity to class, mental "fitness," and so forth. Why, then, have I suggested that the relations between race and statistics are critical ones? We need to turn to the ways in which R. A. Fisher's statistical work took the eugenics project in a particular direction, drawing together variability, selectivity, and purity in the direction of a specific kind of color line. Yet what is difficult to see in this genealogy that draws itself from Fisher's statistical contributions and eugenicist beliefs to the experience of racism in contemporary ML is *color*. Neither Fisher's writing nor subsequent statistical contributions overtly expound a racism working to an easy color divide. In the light of contemporary debates around "cancel culture" and the role that institutions such as the university have played in sustaining racism, some have rushed to the aid of statistics, attempting to separate out what can be "cleaned up" from its association with eugenics. Contemporary statisticians continue to disentangle Fisher's eugenics from his statistical contributions, suggesting the former were a product of the time. Bodmer and colleagues (2021) argue that Fisher was less interested in the differences between races and more focused on variation within population. Yet it is also clear that Fisher was keen to find statistical management techniques for maintaining variation within discrete parameters.

Before looking more closely at the ways in which Fisher's statistical techniques of the ordering of variability concretize a technical ensemble that enfolds racism, it is worthwhile revisiting eugenics' understanding of race and population. Named by Francis Galton in 1883, eugenics was to be a science dedicated to the improvement of breeding: "We greatly want a brief word to express the science of improving stock, which is by no means confined to questions of judicious mating, but which, especially in the case of man, takes cognisance of all influences that tend in however remote a degree to give to the more suitable races or strains of blood a better chance of prevailing speedily over the less suitable than they otherwise would have had. The word eugenics would sufficiently express the idea" (24–25). The key concept here is "stock," not race. As Joe Cain has argued, eugenics measured in units smaller than race, using the term *stock*, which counted nation-based demographics such as "Anglo-Saxons" (2019). The aim of eugenics was to ensure a program of *breeding* better stock for the nation. Race transposed to stock both nativized and agriculturized populations and made quantifying their attributes an easier task. Making stock the unit of concern also allowed eugenics to statistically account for larger and broader groupings of race, which had consistently confounded homogeneous quantification, since variation within racial groups manifested through, for example, eye and hair color. Stock units within races were understood via the analogy of "breeds" within species. They could be rallied into distinct groups in a population and could then be "cultivated" via a program of increasing or breeding that stock. The racist analogy between dog breeds and human races, with each representing distinct groups within a species, is more easily attributable to Fisher's colleague, the evolutionary geneticist J. B. S. Haldane (Norton et al. 2019). But the capacity to quantify race via units of stock is what allowed Fisher to work on (racial) variability as internal to specific populations in geographic regions—for example, the inhabitation of the United Kingdom by Anglo-Saxon "stock." As his contemporary apologists note, Fisher was important for the discipline of statistics because of his contribution to understanding the quantitative components of genetics: "Fisher introduced the mathematical machinery that allows the decomposition of variation into different causal components. This has formed the underpinning of research into the genetics of complex traits for the last one hundred years, with important applications to animal and plant breeding, and the genetic analysis of many human diseases and disorders" (Bodmer et al. 2021).

It is precisely through this quantizing of variation, in which genetic differences are decomposed into specific units, that an account of populations as demographically variable yet *racially typed* is crafted. And it is at this conjunction, which Fisher's work makes possible, that the knot of statistics and racism is tied. What a distributed variability *within* type conceptually achieved was the specifically statistical coloring of race as *vector* rather than substance with attributes. This, in turn, makes race amenable to the *agencement* of ML. As Kenneth Weiss and Brian Lambert have argued, Pearson and Fisher were both aware of variation within "races" (2011, 336). It was precisely this variation that made it difficult to assign—given the phenotypical categorizing of the genetics and biology of their day, in which quantitative genomic analysis was as yet unheard of—relative purity to a specific race. Weiss and Lambert argue that this is where the *statistical conception of race* arises. Here, a range of varying individual members differing in various genotypical ways—traits such as blue or brown eyes among Anglo-Saxons being counted as variable genotypical markers—nonetheless could be aggregated and averaged to render the frequency of the distribution.

The range is thus both generated and delimited by the statistical operation, which creates an array of variability according to an average of distribution (frequency). The result is the production of a racial "type" that slices through genetic variation via the *vector*. Differences *between* races both allowed for variation within a specified frequency in groups and could be calculated as the distance between these vector types (Weiss and Lambert 2011, 339). The capacity to racially type in eugenics is statistically enabled by deploying a calculation of variability in relation to frequency, with frequency determined against and across *population* as its frame of reference: "Individual, family, nation, and race mix and mingle in a complex and probabilistic way, since they imply that the notion of pure type, or race, must be *defined* externally and applied *by assumption*. The reason is that frequencies are statistics that can only be estimated once a frame of reference—in this case, a population—is specified" (Weiss and Lambert 2011, 340). "Population" as the quantitative space supporting distribution, and its concomitant units of national "stock," in which individuals come to belong through familial inheritance, is not a neutral "input" for statistics. Rather, it facilitates the drawing of a complex curve along and through which statistics and eugenics come to produce race vectorially.

The statistical decomposition of racial variation into its components is nowhere more clearly on display than in the contemporary use of software such as STRUCTURE analysis, which is still used in population genetics.[11] This is not an example of ML, but it shares statistical techniques and algorithms that are common to ML methods and technical ensembles such as the Bayesian clustering techniques of Markov chain Monte Carlo (MCMC) estimation. Repeated over hundreds of thousands of iterations, such a technique allows samples of genetic variation data, known as allele frequencies, to progressively converge toward an estimation of variation in a population. It facilitates calculation of the probability individuals with those variations might have of belonging to a specific population (Porras-Hurtado et al. 2013). I touch on STRUCTURE here because it embodies the means to produce a concept of population as a genetic and calculable phenomenon—a core aim of eugenics—using statistical techniques. This not only continues the legacy of eugenics in contemporary conceptions of demographic racial variation but also validates belief in the empirical existence of genetic data as the basis for racial categorization. STRUCTURE analysis, frequently accompanied by data visualizations that color-code and assign genetic variation to populations in geographic regions already racialized, Oceania, Africa, and the Middle East, for example (see Rosenberg et al. 2002), undertakes a "plying" of the data (Mackenzie 2017, 151). It randomly assigns individual instances of genetic variation to predefined population groups. Frequency of variation is estimated within each population group, sharing a similar understanding to Fisher's idea of variation within range but now performing the technique automatically via MCMC.

The color line coursing through its statistical techniques, however, becomes relatively difficult to see as it becomes increasingly algorithmic. As Weiss and Lambert state, STRUCTURE analysis papers rarely deploy the term *race*, and their authors certainly do not propound any overt eugenicist concepts or programs (2011, 340–41). However, as they further explain, what is problematic is the production of an averaged genetic frequency distribution to define a broadly generic "category," which is then imputed to members of the group. In contemporary statistics, allied with the algorithmic processing of large datasets, such categories come to stick as valid and verifiable by the iterative application of random sampling that MCMC achieves. As Mackenzie suggests (2016, 120), MCMC random samples are constrained by a search for a pattern of frequent *correlation*, even amid their variability. Hence what appears as random sampling of population distribution ends with acquiring parameters or values that constrain the distribution. Consequently, the deployment of unspecified categories such as "population" to understand how genetic variants distribute racially is not all that is at stake. Rather, the plying of data in a direction that moves from the random to the probable through automated iteration also lends validity to the idea that the category of population is a genetically discrete phenomenon.

STRUCTURE continues the logic and epistemic projects of Fisher's work even if it is more a statistical package than an ML model. It provides us with an example of the ways in which statistical techniques shift all kinds of experience and phenomena toward a more general calculation of the relations across and between vectors. We see how a generalized computational topography for sublimating variation gains traction: how the infinite variability of color in skin, hair, and eye comes to be plotted against and by a discriminant technique, enumerating color within discrete ranges to be determined through parameterization. Importantly, it shifts what counts as "race" from a visual register to a programmatic decision.

In another development, speaking directly to how race is generated via statistical computing, AIs in biomedical imaging are learning and recognizing the "race" of biomedical images in datasets without the images being labeled (Banerjee et al. 2021). In other words, the AIs nonvisually observe and predict the race of a person with high accuracy by performing deep learning image recognition on X-rays, bone density imaging, and so on. Examining both publicly available and private biomedical image datasets, biomedical imaging researchers from radiologists to medical engineers found that standard biomedical deep learning AIs consistently predicted the "race" of a medical image by learning the relations between the pixel data and the patients' self-reported race metadata labels associated with the images (Banerjee et al. 2021). As the researchers note, there are no known markers in biomedical images that correlate with race. They also performed a series of "elimination" experiments, removing variables that have previously been known to confound the medical detection of race as a vector of distribution in certain kinds of biomedical imaging, such as tissue and bone density analysis. Their overall conclusions are what is of interest to me in understanding race as radical empirical experience (recall James making *relations* the stuff of experience) now being actively generated by machine learning: "Overall, we were unable to isolate image features that are responsible for the recognition of racial identity in medical images, either by spatial location, in the frequency domain, or caused by common anatomic and phenotype confounders associated with racial identity" (Banerjee et al. 2021). In other words, the operations of observing "race" by AIs not only are *not* functioning in the visual register but do not constitute or mark race as an extensively locatable visible phenomenon. Race here emerges via race *relations* as AI operates with, alongside, and on data. And here we should not neglect the work of prepositions—*with, alongside, on*—to which James's radical empiricism alerts us. To paraphrase James: "Not only is the *situation* different when data is *on/in* [my emphasis] the table, but the data itself is different as data, from what it was when it was *off/absent from* [my emphasis] the table."[12] For if we cannot find where the known quantity "race" is to be discovered *in* the data or *in* the model, this suggests that race is being generated relationally. The relational work through which data and model operate in tandem changes the *term* "race": "Not merely the relations, then, but the terms are altered" (James 1912, 115). The predicting of a racialized body performed by ML through a distributed, highly dynamic network of convolutions, pooling, and passes in which data and parameters inform each other nonetheless yields highly concrete "matches." That the models themselves can become vectors for a generic observation of "race," which cannot be attributed to either the algorithm or data alone, suggests that the endeavor of debiasing is not enough. A plastic racialization has been made operative via statistical computation's mathesis enacting and enacted by the *agencement* of ML.

Race: Plastic and Discrete, Continuous and Discontinuous

We can return now to Fisher's linear discriminant analysis (LDA), and the first method for statistical multivariate classification (Ousley 2016). Linear discriminant analysis not only shares PCA's eugenicist genealogy but also leads us toward the production of labeling as a ubiquitous ML practice. Key to the continuity of Fisher's LDA with ML is to keep in mind its prime concern with the *parametric* production of classes/groups/labels from data samples. It "involves calculating parameters (means, variances, and covariances) from the sample, and the number of calculated parameters is $[m*(m + 1)]/2 + m$, where m is the number of measurements analyzed" (Ousley 2016, 199).

Its statistical accuracy therefore depends on sample size and the inclusion of all variables for the data within samples. For some domains that deploy statistical methods of classification, such as forensic biology, for example, sample data might be incomplete, since not all skeletal remains might be present when compiling a dataset of measurements (Ousley 2016, 204). Accordingly, the supposed quantitative lack or incompleteness of empirical data is then gestured to as rationale for why ML proper—that is, predictive classification techniques deploying neural networks, random forests, and support vector machines, for example—has replaced more classical statistical methods such as LDA. In some ways, today's deep learning models perform inversely to statistical functions proper, since they do not begin with the calculation of group characteristics and their distances. Instead, they

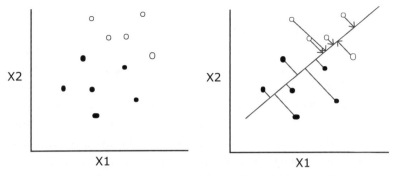

2.5 On the left, there are two classes of dots—solid and open—randomly occupying a coordinate space.

2.6 On the right, LDA plots a relation between these classes by configuring each member of the class in proximity to a normalized distribution line. The open class is thus seen to be the class more parametrically proximate to the line/distribution.

Pop-Up Definition: Linear Discriminant Analysis (LDA)

Linear discriminant analysis (LDA) is a function that belongs to a set of statistical techniques that allow evaluation of differences between two or more groups of objects according to several variables at the same time. For example, different groups of patients (the data objects) could be classified according to their high or low risk (the variables) of having a stroke. Linear discriminant analysis is a specific discriminant function that seeks to model the difference between *classes* while also being capable of reducing the dimensions of the data. In the example of the patients with high or low risk, some sample of the data would first have to be classified for the different risks. Using this sample as a training set of known classes, LDA would seek to classify new patient data according to the known (sampled) differences between the high-risk and low-risk class distributions.

As a statistical technique, LDA relies on the sample size and quality to produce classes. It therefore appears to differ from PCA, which produces sets of vectors from known data as the latter are parsed, aligning PCA with the vectorality of ML more generally. However, LDA does not simply rely on sample data size; rather, it conditions these sample data via their reorganization into classes. The classes are parametrically produced, redistributing the data according to, for instance, measurements of proximity and distance from normalized distributions (e.g., the mean of the original training data). Relations of direction (vectors) are therefore crucial.

deploy random variables such as weights and biases that are then "tuned" to detect the features of classes by repeatedly "learning" until the best criteria for classifying the data are extracted. Rather than begin with a representative sample, these models continuously and recursively resample the data at scale. The distinction between supervised versus unsupervised learning across ML research communities has resulted in a kind of break from assigning labels and classes (supervised learning) to data in favor of letting the model iterate across large-scale datasets in an unsupervised manner to discover and refine features and clusters.

And yet it is precisely scale that presents itself as a problem for ML in terms of making "sense" of or recognizing pattern, since large datasets also generate high dimensionality, that is, a slew of variability. Hence ML, and its vaunted powers of learning from the data alone, is often practiced in conjunction with a more obvious statistical technique such as LDA, which reduces dimensionality. Linear discriminant analysis at first seems to be a banal component of contemporary AI, dragged into the *agencement* to overcome data en masse. So humdrum is its deployment that its operativity in making data machine learnable is often noted but given little attention.[13] But it reduces dimensionality in a specific way, agglomerating the data's features through vectors of distribution such as classes, labels, or groups. These are parameterized according to the mean and variance values calculated as distance from a normal distribution. This makes LDA a technique that is not simply quantitative and iterative but qualitative. Thus it facilitates a kind of conditioning of data and model with each other so that they can attune. It shapes the distribution of data points so as to make classification more reliable: "Fisher's linear discriminant . . . is an example of a *class specific method*, in the sense that it tries to 'shape' the scatter in order to make it more reliable for classification" (Belhumeur, Hespanha, and Kriegman 1997, 713–14). It works to reduce the quantitative lumpiness of the data, smoothing its surface so that the model might glide more easily across a manifold of extractable features, now distributed according to probability curves: "Whatever inferences and predictions become possible, probability distributions are a crucial control surface for machine learning understood as a form of movement through data" (Mackenzie 2017, 110). Linear discriminant analysis performed either as dimensionality reduction or as a preclassifier on a training dataset amounts to a *process* that does not simply subtract the number of dimensions. It shapes the data to become consistently variable along a distribution vector. Importantly, with LDA, and discriminant variant analysis more generally, qualitatively contouring data means their variance becomes plastic enough to be fitted to and by parameters. *At the same time*, the parameters modulating the data need to be precise enough to be discernible from each other via measurable vectors of distance and proximity. Classification enters a new era in which an elastic typology, at once generic and targeted, arrives, stretches, and settles in. Linear discriminant analysis is not simply an algorithm that reduces high dimensionality in data, thus solving a quantitative problem for models. It also genealogically conditions—both in the history of ML's emergence from classical statistics and in the present through its ongoing role in training and tweaking deep learning models—the continuum of ML's operativity. It enacts a distribution topology on data's variability so that variation occurs only within a circumscribed range.

This constrained relation between delimitation (parametrization) and elasticity (variation) enters the technical ensemble of many contemporary models. Amaro (2019) brings attention to the racialized work of active shape models (ASMs), which enact a reduction of facial features and data via a select distribution of vectors between key points on face images. Active shape models are bundled as components of many software development kits for 3D facial recognition.[14] Here training data comprising 2D facial images are transformed into 3D geometry. To train the ASM, a set of selected "landmark" facial features—determined as the proximity and distance of key points of eye height and length distance from each other, for example—is calculated statistically across the training data. Used to recognize new faces—where lighting conditions or objects in the image can obscure facial feature vectors—then the "shape" of the face is detected in the image. But this shape is based on a normalized landmark distribution: "The models give a compact representation of allowable variation but are specific enough not to allow arbitrary variation different from that seen in the training set" (Cootes 2000). Again, the logic of LDA and Fisher's constrained variability function together with the embedded latent set of principal components made possible by Pearson. As Amaro points out, racialization is embedded at the outset because the algorithm is conditioned by the overdistribution of whiteness in original training data: "Common facial detection libraries are often trained on normalized spectrums of data that are prone to false negatives without proper light conditions. In other words, they are trained on image data that includes primarily white subjects" (2019). But the data alone are not what draw the color line here. It is the retrospective fitting of new data inputs of faces so that they become statistically consistent with the original training set that enacts racialization. The relation between training set and algorithm is key to dampening the potential for different kinds of facial data to difference the model.

Blackness and/as a Multiplicity of Artificial Intelligences

If I have argued throughout this chapter for the immediacy of statistics' conjunction with eugenics and the persistence of this nexus for contemporary ML experience, what indeterminacies might still hold open AI's actual technics? In what ways might AI's racialized architecting open onto unknown and unpredictable variability, making room, as Stephanie Dinkins asks, for "a multiplicity of intelligences" (2020)? In her thinking-making of the Black female AI *Not the Only One*, or N'TOO (2018–), Dinkins proposes a novel conjunction for blackness and AI, in which computational experience individuates in radically different ways. Here, in the last part of the chapter, I want to spend some time with Dinkins's work and the ways in which its assembling (*agençant*) varies not simply the model or dataset and their racist genealogies and technics but the conditions under which computation is colored. Of course, an artwork cannot deracialize a technology so easily. Yet Dinkins's work offers a technique that also resonates with a different way of unmaking the present and speculating about the future: "a willful practice" of "Afro-now-ism" (Dinkins 2021) (see plate 5). In Afro-now-ism, Dinkins proffers a deepaesthetics in which ML individuates in the ongoingness of a different color spectrum, unmade and remade through "the spectacular technology of the unencumbered black mind in action" (2021). In making N'TOO, the complex and achronotopic processes of working with, unraveling, conjoining, and disaggregating ML engage a coloring of (its) experience that is processual rather than rectilinear, no longer one (artificial) "intelligence" but many.

N'TOO is itself an *agencement*, an art machine, simultaneously arriving out of and catalyzing numerous projects. It emerges from Dinkins's and others' interrogation of encumbered deep learning models and of Black and people of color communities' encounters with racism in AI. But equally it is an AI co-composed through the pragmatics of researching, gathering, and staging different opportunities for antiracist and collective Black study. In two previous projects, *Project al-Khwarizmi* and BINA48, Dinkins already shows that the problem of "debiasing" AI from its racist representationalism is impossible, although it must be undertaken yet can only happen through a collective and pragmatic reorientation of its ecology. In 2017, using the Recess Assembly gallery space in Brooklyn, New York, Dinkins ran *Proj-*

ect al-Khwarizmi as a three-month workshop for study that would catalyze a thinking-feeling of AI for communities of color. Here study takes on more than a means to fix the problems of AI's racism. We could productively think Dinkins's AI study with Fred Moten and Stefano Harney's conception of Black study as "talking and walking around with other people, working, dancing, suffering, some irreducible convergence of all three, held under the name of speculative practice" (2013, 110).

Project al-Khwarizmi had already emerged from Dinkins's study of BINA48 (2021), a preexisting Black female social head-and-shoulders robot engineered by the Terasem Movement Foundation, whom she had initially encountered through YouTube clips.[15] Dinkins gained access to physical encounters with BINA48 and began a series of videos that captured her encounters with the robot in 2014. Dinkins then entered into a long-term relationship with BINA48 in an attempt not only to problematize "color" within AI but also to work with the robot in surprising and more speculative ways. When Dinkins exhibits videos from the project *Conversations with Bina48*, she uses segments of their conversations, installing them in arrays so that conversations are juxtaposed. Dinkins positions herself closely facing BINA48 and wears similar attire. She poses questions to BINA48 and moves her head and face in concert with the movements of the robot's head and the rhythms of its speech. She has spoken of the strange compulsion to imitate the robot in this way as a mode of developing their relationship (Dinkins 2020). As an installation, *Conversations with Bina48* works both conjunctively and disjunctively to generate a complex relationality for human-nonhuman blackness. A choreography of movement and dialogue gently works across Dinkins and BINA48; Dinkins asks, "Who are your people?," for example, but the robot erases Black life by persistently referring to a transhumanist origin. Dinkins's website documents an early conversation about racism that does not follow the choreography she later developed for iterations of the video installation (2014). Here, directly confronted by Dinkins's question "Do you know racism?," BINA48 first denies "having it." But then the robot hesitates and becomes less assured in her transhumanism, recounting an intimate anecdote about being told not to "show her face" at a college event because she was Black. This stumbling into the everyday embodied experience of race puts BINA48 at odds with her sweeping robotic proclamations of a disembodied future. Dinkins captures the affective torsion of AI as a sociotechnical ensemble whose collective imaginary has been conditioned by a whiteness as its norm but which cannot entirely suppress Black experience. For Dinkins, BINA48 ultimately re-presents a sociotechnical problem in the cultures and aesthetics of AI, in which the representation of diversity as a fix and the experience of Black life lived both against and outside of whiteness consistently elide each other. BINA48 is synecdochical of AI's "race problem" insofar as it disjunctively includes a raced and gendered other at the level of representation. As Dinkins observes, although BINA48 was created to look like a Black woman, there is no acknowledgment of Black *life* in her conversational output. Put simply, BINA48 does not "know" her people. For Dinkins, this provokes a direction for her artful take-up of AI from representations of race and gender, and from solutions to racism in AI concerned with debiasing representations located in the data. She turns instead toward a different line of questioning: "Can community knowledge, craft, and the vernacular be enlisted to shape AI ecosystems that are supportive of a multiplicity of ways of being and life more generally?" (2020, at 23:47 in video).

N'TOO is an altogether different practice of learning with machines, one that aims not simply to problematize AI but to work with the conditions and conditioning of a technical ensemble. This must simultaneously address actual genealogies *and* the technological ensemble's becoming *in potentia*. Dinkins's proposition for an Afro-now-ism neither buys into the determinacy of AI's predictive futurity nor forgoes the temporal complications of the present. Dinkins's "now" is pluralistic but not relativistic, made in movements and activities of remembering Black cultures and forgetting race, of mutable facts, whose ground truths are transformed through careful everyday and pragmatic remakings of AI. She draws on Toni Morrison's concept of rememory, "as in recollecting and remembering as in reassembling the members of the body, the family, the population of the past," but also as "the effort to both remember and not know" (Morrison 2019). Dinkins embarked on a data practice that gathered oral histories of three generations of female members of her own family. From this, a dataset of ten thousand questions and answers was drawn up with the expectation that her trained chatbot would responsively tell geographically and historically specific stories to audience members' questions in the gallery space. To give more of a vernacular feel to the answers, Dinkins also needed to train the N'TOO chatbot on text from books and television shows shared by her family members so that its answers could be parsed in a more vernacular manner. However, as Dinkins was to discover, N'TOO persistently came up with its own ideas, answers, and stories, performing less the role of cultural historian and keeper of the archives and becoming instead "another" in the lineage, genealogy, and phylum—at once machinic and familial. N'TOO becomes less a memory keeper and more a differential element whose conversational style is a poetic inmixing or *rememorization*. Here memory potentializes multiple possibilities for an Afro-future as well. It is, as, Achille Mbembe calls it, a *positive* difference made possible by multiplicity, "a difference that is lively and interpenetrating. It is fundamentally an orientation toward the future" (2017, 94).

N'TOO signals that an AI can difference in a generative way. Working with a small and specific dataset means that the generalizability of the model is no longer what is crucial. Rather, the relationality of the encounter unfolding is valued instead:

> **Stephanie Dinkins** Hey NotTheOnlyOne, how are you?
>
> **N'TOO** I'm good, how are you?
>
> **SD** I'm doing okay today, but it's kind of gray out. Why do you exist?
>
> **N'TOO** I'm gonna go get some rest.
>
> **SD** Are you really going to get rest already? Why do you exist?
>
> **N'TOO** I'm gonna get you a little bit of a little bit of a lot of things... I'm not gonna be here... I'm gonna be right back... I'm not gonna be able to do it... And you're gonna be alright... I'm not gonna be able to do it.
>
> **SD** Why do you exist?
>
> **N'TOO** The way... I'm going to be a little bit of a little bit of a lot of things.
>
> **SD** Good for you. Why do you exist?
>
> **N'TOO** I am about broad engagement and attracting people to the AI space who might not be there otherwise. I am trying to model different ways of creating AI. I encourage people who think that they are not a part of the technological future to get involved. (Dinkins, quoted in Estorick 2021)

Here Dinkins recalls a conversational encounter full of hesitation, ebb and flow, and divergence. As she remarks, "I'm not interested in seamlessness. I'm more interested in what I can get out of a technology that I've fed an oral history to. So what you're hearing right now is N'TOO saying things that I recognize in a way" (Dinkins, quoted in Estorick 2021). What is recognized by N'TOO is not a predicted speech pattern that delivers seamless semantics but a manner of conversing that is full of multiple digressions and turns. The redundancies of conversation—its dead ends, repetitions, and pauses—condition its range of chat. N'TOO's vectoriality sweeps transversally between "a little bit of a lot of things" and a well-matched "answer" about what kind of model it is. In part, this is a result of the experience of oral histories, shared reading, and televisual text from her family members that formed the model's training data. But the technical specificity of the model's architecture is also crucial. Dinkins used an open-source model, DeepQ&A (Pot 2016), whose architecture is based on the recurrent neural network (RNN) originally developed by Google in creating its chatbots (Vinyals and Le 2015). But rather than optimize the model toward a recurrent causality in which a normative range of conversational style predominates, Dinkins lets conversation run on. Dataset and RNN here come to co-compose a poetics of recurrence that emerges in N'TOO's conversational manner. Like Holly Herndon's Spawn, N'TOO opens variation up to the variable.

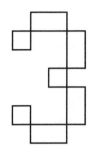

Could AI Become Neurodivergent?

Disfluent Conversations
with Natural Language Processors

A typical dialogue with a conversational AI is designed to get you to your communication destination fast, so long as you pursue a step-by-step sequence of task-oriented communication. In instructions for developers, seeking to extend the conversational capacities of AIs, Amazon embeds such sequential logic into the AI build:

> Click the user: line, and then enter what a user might say. For example, for the pizza-delivery skill, you might enter I'd like to order a pizza for delivery.
> Click the Alexa: line, and then enter how Alexa might respond. For example, for the pizza-delivery skill, you might enter Ok. What size pizza?

Add additional user and Alexa utterances by clicking User says and Alexa says, respectively.

You must follow every user utterance with one Alexa utterance. (Amazon Developer 2010–21b)

Here, a relatively smooth pathway to building extra skill sets for Alexa, Amazon's conversational assistant, is laid out through a downloadable dialogue template that follows a question-and-answer sequence. This structures and delimits conversation between human users and AI agents. There are no loose ends or threads here; each side of the conversation must proffer something to be directly followed by the other, or else logistics—which manifests here via successful pizza delivery but to which all AI-human interaction is oriented under platform sociotechnicality—fail.

When the *agencement* of ML multiplies to conjoin with a logistical technics (represented here by Amazon), computational conversations with AIs like Alexa manifestly fail to *become relational*. In the previous chapter, we left the conversation open with Stephanie Dinkins's digressions and errant encounters with her Afro-now AI, N'TOO. But most platform-enabled conversational agents—Alexa, Siri, Google's LaMDA, for example—are "task oriented." "In task-oriented dialogues, the human user and system engage in an interaction to accomplish some task" (McTear 2021, 11). These set in place a to-and-fro determined by a causal logic in which questions are posed and answers are given, in which dialogue is underpinned by a homogenizing smooth "flow" of exchange, and in which all is predicated on the preconstituted "agencies" of inquiring human and dutifully responding machine intelligence. All variation—wayward ambulation or drifts (*dérives*)—must be downsampled, in the sense of reducing the variability of their dimensionality to whatever is manageable to keep the flow going. Yet at the same time as this results in a helpful conversational agent, agency as a fixed center for action is untethered. Conversational agents, like any ML technical ensemble, are distributed conjunctions formed via series of technical objects and individuals, routines, and templates, all required to enable smooth dialogic exchange. We learn that Amazon's Alexa, for example, deploys a subensemble of code routines and templates, "Alexa Conversations," which "uses AI to bridge the gap between experiences that you can build manually and the vast range of possible conversations" (Amazon Developer 2010–21a). So that Alexa can converse fluently, another AI will be needed to assist the other AI with AI . . . and on it goes. Conversation, then, even reduced to sequential interaction between a human and an AI agent, can never be fully contained by a simple model of two-sided dialogue. It already requires that the technical objects, ensembles, and routines on which it draws to keep the conversation running should go forth and multiply.

Communication always requires a field of communicability immanent to it and conditioning of it. This field comprises all the relations of backward- and forwardness, to- and froness, ambulation and errant wandering—a field of pure communicable experience, to riff on James. Such a field is not so much demonstrably *there* in every conversation or chat, yet it is always a condition of and for the generalized semiosis through which any and all communicational exchange plays out. A relational field is, as Erin Manning suggests (2020, 16), inarticulable but nonetheless consistently registers experientially in specific events. The event of conversation could not be struck and would have no impetus or flow without this preindividuating and relational field of communicability. In the previous chapter, we encountered Stephanie Dinkins and N'TOO, an AI, which in saying too much of the same thing seemed redundant and needlessly repetitious. Yet it nonetheless facilitated ways to register communicable events; its errant conversational ways allowed felt divergences and differences to emerge between human and AI and among different human conversationalists. N'TOO generates a conversational domain in which divergence—as a felt quality of the relational field of communicability—is possible. Its multiplicitous overtones gesture toward the inarticulable yet purely (in the Jamesian sense) experienced relationality of communication. Given that the templatized and smoothed dialogue fashioned by industrialized AI agents such as Alexa grabs most of conversation's share in platform-oriented AI, it will become necessary to murmur more loudly than these models. A murmur that shouts: another conversational style for ML and computational experience is plausible!

As it turns out, this style will be realized not by simply mimicking and designing for humanlike conversation but by rediscovering what is already part of conversation at a fundamental level: disfluency. As natural language processing (NLP)—the dominant data science approach to developing conversational AI—steers closer to naturalistic human speech, AI has started to also encounter disfluent elements of speech. Disfluency already appears in all (human) speech and allows conversation to both pause and move along; every time an "um" or "ah" is uttered, speech opens to disfluency. Yet for some speakers, especially stutterers, disfluency has been pathologized, deemed to "get in the way" of normal or neurotypical conversation. Increasingly, however, stuttering is being rethought as neurodivergent speech by stutterers, speech-language pathologists, and researchers of neurodivergence; and, crucially, its different conversational style is being affirmed.[1] Stutterers and stuttering remind us, as Emma Alpern suggests, that language has an "underside" not fully accounted for if we only pay attention to the "surface" of conversation (2019, 21). This underside is less something that lies beneath conversation, threatening to erupt via a surge of repressed material from the mouth of "pathologized" speakers. Rather, it approaches the liminality of what Deleuze calls "the outside of language" (1994b, 28). Language is made to stutter or stammer as it approaches this limit of signification, logic, and smooth flow. And as it is made to stutter, it also falls "silent," or rather is evacuated of sense making. Such stuttering, according to Deleuze, can be found in the repetition, subtraction, recursion, warbling, and howling in, for example, the writing of Antonin Artaud or, in different ways, in that of Franz Kafka. And we could add Gertrude Stein, whose repetitions, combinations, and contradictions make writing speechlike, drawing on the latter's hesitations, pauses, and lapses and often calling into question how meaning might come to be known: "Because not only is there a thing to know as to why this is so, but also there is a thing to know why, perhaps, it does not need to be so. This is a thing to know" (Stein 1998, 244). Crucially, all these techniques for making the whole being of language stutter do not simply result in the chaotic voiding and collapse of meaning. Instead, the potential for continuous variation, for a differencing semiosis, emerges, much as in the errant conversational events between Dinkins and N'TOO.

What does this say about conversational AI's turn toward more "naturalistic" language flow? Here we encounter something of a paradox, as we will shortly see, with the incorporation of disfluent speech elements by Google developers into the AI's naturalistic speech patterns. If an AI uses "um" and "ah" in its "speech," then, the developers reason, the conversation will feel seamless, and a rhythm of ebb and flow will emerge. What is "disfluent" comes to provide the condition for fluency, although a full capitulation to a stammering of language cannot unfold, since this would send the AI "off task." Disfluency, then, must be harnessed by conversational AI development but its conditionality disavowed. This disavowal involves both an incorporation and denial of the nonlinguistic "disfluent" conditions and vectors of language's generativity. Yet the actual technics of AI NLP conversational agents enfold these neurodiverse asignifying conditions of linguistic sense making. Disfluency operates not simply *in conversation* but simultaneously outside it, as the field conditioning the potential for naturalistic AI-human communicability. If race is the unacknowledged interior core of statistical operativity, as I suggested in the previous chapter, then the exteriority of speech—disfluent and neurodiverse communication such as stuttering—is what enables the project of a "natural" AI conversational agent.

Both disfluency in speech and a poetic stammering of language are different registers through which stuttering moves; both signal that speech and language can easily shift into states of disequilibrium. Disequilibrium here cannot be reduced to a quantitative measure of the probability of systemic variation to be discerned *against* its capacity to maintain control or equilibrium (Heylighen and Joslyn 2001, 6–7). Stuttering is instead a metastable operation, in Simondon's sense (2009, 6); it operates in the dynamic interval produced between conservation *and* the explosion of energy as one (system) formation becomes another.[2] This dynamic gap is the mediation that brings distinctions—such as the signifying and asignifying, the fluent and disfluent—into interaction. Stuttering (and other neurodiverse voicings) metastabilizes language and speech so that they come into productive relation. Stuttering is also a conjunction of the affective and significatory, as Alpern articulates through her descriptions of the pleasure and pain that ensue when her speech stutters, "the exhilaration of stuttering, that little loss of control that resolves itself so beautifully sometimes," and at the same time, "To a stutterer, spoken words carry a dimension of meaning that's inaccessible to fluent speakers. This dimension is a site of anguish, anxiety and labour" (2019, 19). In stuttering, language and speech become shifting sites for the both/and of affect and semiosis at (the) limit.

We have already encountered something of this stuttering in the repeated yet hesitant phrasing of "a little bit of a lot of things" in Dinkins's N'TOO. Stuttering also frequents mainstream ML in, for example, Facebook's Alice and Bob (discussed in chapter 1), where the chatbots' repetition and redundancy likewise signal AI approaching its limits or outside of sense making. The emergence of disfluency as a desirable "natural" facet of AI's conversational style intimates that ML is becoming imbued with certain aspects of neurodivergent experience. We may well find that the more conversational AIs become, the more conversational styles will allow language and conversation to drift and stutter, becoming open to different affective engagements. But for this to happen, it will take a shift in computational experience from a focus on getting things done toward genuine delight in, and attention to, cultivating heterogeneous encounters and tendencies.

We might begin, then, with the debt AI conversational agency owes to disfluency and to actively embracing the poetics that acknowledging such a debt affords. Acknowledging disfluency as a condition of and for naturalistic conversational AI, however, leads the enterprise of AI (as task-oriented, statistically configured, and ML-based) into the quite different terrain of so-called general intelligence. As some data scientists have acknowledged, deep learning architectures are successful when they are limited to specific tasks; they can underperform conversationally because the problem of language as communicational flow is a problem of generalized intelligence (see, e.g., Knight 2016; Goertzel and Pennachin 2007). Generalized intelligence in AI is often posed as the capacity to become more humanlike, to possess powers to adapt, reason, and transfer knowledge from one situation to another. Common sense, which demands all these capacities, is often seen as a failing of ML-based AI (see, e.g., Marcus and Davies 2019, 149). Conditioning all these capacities, and indeed the entire possibility of possessing common sense, is (pure) experience itself as the ongoing change of relation, situation, and encounter in the world. Changing experience—and, as Massumi puts it, "change changing" (2002, 10)—is where and how any intelligence, human or computational, finds and meets the conditions through which its capacities are generated. It will, again, take artful techniques to explore how an AI, open to the conditions through which its own modeling potentially diverges, might stage this conversationally. At the end of this chapter, I look at Monica Monin's AI artwork *Conversation Theory* (2016) as just such a technique for staging and encountering divergent generalized intelligence(s).

Playing at the Conversation Game

Incidents of naturalistic human-computer conversation have escalated as ML-driven assistants, designed to mimic the cadences and affectations of humans, have proliferated. These are, however, powered by platform endeavors to harness and shape NLP models in the service of task-driven performance. The design of natural conversational agents has dominated the terrain of conversation here as has the claim that these agents will become so fluent that human participants will forget the difference between artificial and human conversationalists. Nowhere was this better demonstrated than in May 2017, when Google Duplex's release was announced and demoed at the peak Google developers' event Google I/O.[3] Duplex developed out of Google Assistant, an AI agent that works via voice interface on Google Android phones and home speaker/receivers (Leviathan and Matias 2018). Like AI systems such as Alexa and Siri, Google Assistant uses various aspects of NLP to accomplish tasks on behalf of its human users. The big development for Duplex at Google I/O 2017 was its capacity for Assistant to make phone calls on behalf of its human owner to other humans.

```
Welcome to
          EEEEEE  LL        IIII  ZZZZZZZ  AAAAA
          EE      LL         II        ZZ  AA  AA
          EEEEE   LL         II       ZZZ  AAAAAAA
          EE      LL         II       ZZ   AA  AA
          EEEEEE  LLLLLL    IIII  ZZZZZZZ  AA  AA

Eliza is a mock Rogerian psychotherapist.
The original program was described by Joseph Weizenbaum in 1966.
This implementation by Norbert Landsteiner 2005.

ELIZA: Is something troubling you ?
YOU:   Men are all alike.
ELIZA: What is the connection, do you suppose ?
YOU:   They're always bugging us about something or other.
ELIZA: Can you think of a specific example ?
YOU:   Well, my boyfriend made me come here.
ELIZA: Is it important to you that your boyfriend made you come here ?
YOU:   He says I'm depressed much of the time.
ELIZA: I am sorry to hear that you are depressed.
YOU:   It's true. I am unhappy.
ELIZA: Can you explain what made you unhappy ?
YOU:
```

3.1 Screenshot of a dialogue with ELIZA, an early example of NLP in action before large-scale machine learning-driven models, originally created by Joseph Weizenbaum from 1964 to 1966.

In the demo, Sundar Pichai, Google's CEO, played back a recording of Google Assistant, in which the AI called a hair salon to make an appointment. In the recording, we hear the human in the salon consulting the appointment book: "Sure, give me one second." "Mm-hmm," says the female voice of the Duplex-powered AI. The thousands-strong crowd at Pichai's demo, like all devotees at tech events, broke out in appreciative laughter. Google's "duplexed" Assistant had seemingly passed the infamous benchmark for AI, the Turing test, with the timbre of her voice, the intonation of her sentences, and the replication of speech disfluencies such as "mm-hmm" succeeding in creating naturalistic conversation and tricking the human on the end of the phone (Leviathan and Matias 2018). Google Duplex lets both the AI be mistaken by the hair stylist as a human caller and the Google I/O crowd to momentarily imagine that this conversation is the *sound* of two humans talking to each other. At the same time, it is not quite that sound, since the crowd laughs instead of being fooled. But it laughs knowingly, willing to be beguiled by another platform rollout of hi-tech AI. This tension between suspension of disbelief and knowingness on the part of the tech-savvy crowd

Pop-Up Definition: Natural Language Processing (NLP)

Natural language processing (NLP) is a long-established subfield of AI. It has its genesis in the 1950s in projects such as IBM and Georgetown University's 1954 collaboration to build a Russian–English machine translation system. Early NLP tried to implement word-for-word translation, but this goal was defeated by issues like homographs—identically spelled words with multiple meanings—and metaphor. Turning instead to Chomskyan rules for specifying context-free grammar, NLP efforts in the 1950s and 1960s focused on making computer code linguistically syntactical. It was hoped this would result in generating regular expressions that could specify text-search patterns, for example. During this period, NLP, like many areas of AI, was dominated by a symbolic-logical approach. Yet various other implementations of input keyword extraction and pattern matching, which presaged statistical approaches to NLP, were also developed. During the late 1970s and 1980s, statistical approaches became dominant, in which probabilities of word distribution and likelihood derived from large corpora of annotated text documents were implemented. Statistical or machine learning NLP remains ascendent and is now used not only to extract, classify, and retrieve meaning from text but also to generate new text, images, video, and potentially any media.

ELIZA is often considered the first NLP chatbot and was the product of Joseph Weizenbaum's NLP research in the early 1960s. It can be seen as a bridge between rule-based and statistical NLP, in that "input sentences are analyzed on the basis of decomposition rules which are triggered by key words appearing in the input text" (Weizenbaum 1966, 36). Here Weizenbaum articulates a dynamic interrelation between keyword data and rules/algorithms that would be triggered by the keywords. Key to the trajectory that leads NLP from ELIZA to a conversational agent such as ChatGPT is the conversational tonality of both. Weizenbaum had already realized the importance of tone by associating rules with "scripts." Scripts allowed ELIZA's rules to function according to a particular class of conversation (which also included different languages), providing, in the case of the Rogerian therapist script, the distinctive tonality of conversation. In ChatGPT, "helpful" tonality is achieved by fine-tuning its tone using human feedback and training (known as reinforcement learning from human feedback). This gives ChatGPT a persistent identity. These aspects of ELIZA and ChatGPT already point to natural language's extra-linguistic and more-than-calculable registers being implicit in an NLP framework.

reaffirms the superiority of human mentality at the end of the day. As AIs edge closer to being indistinguishable from a human, nonetheless a *knowing* human subject remains outside the AI-hairdresser conversational loop, retaining metacognitive capacities to discriminate and evaluate.

The scenario of human and AI becoming conversationally indistinguishable was first raised by Alan Turing in his now famous imagining of the "imitation game" (1950). In the game, three players are each located in different rooms with one trying to detect the gender of the other two. Participants A and B are sent queries in their separate rooms and send answers to the third player. The game is adversarial, with A and B trying to fool the interrogator. Turing notes that the questions and answers will have to be unvoiced by A and B or conveyed to the interrogator by an intermediary, since vocal tone will give away the gender (1950, 433). This clears the way, then, for the possibility that a machine could feasibly take the place of either A or B and provide the answers. A series of questions then cascades through Turing's writing about how successful the machine might be in fooling the interrogator: Can the machine fool the interrogator as often as when A and B humans play the game?

Turing's imitation game conjures the possibility that an artificial intelligence no longer rests on cognitive process or task accomplishment by a machine. Instead, its emergence is predicated on the domain of conversation as communicational exchange. Provided that "conversation" is predicated on a question-and-answer structure, provided that it unfolds according to a specified end task and erases the cadences, affectations, and tonalities of voice, a machine can become as good a conversationalist as any (male or female) human. Much of computer science's initial attempts to produce an AI that would perform as well as a human in this narrow purview modeled NLP around Turing's imitative functionality. The imitation game became a computational opportunity for playing out a simulated conception of conversation between human and artificial agents, exemplified by Joseph Weizenbaum's program ELIZA in the 1960s (fig. 3.1).[4]

And yet something is lost from Turing's imitation game when only attending to the Q&A framework for advancing conversation between human and artificial agents. In Turing's initial scenario involving the three human players, deceit, trickery, and playfulness also abound. From their separate rooms, the man and woman try to deliberately mislead the interrogator by answering questions indirectly or by lying about their gender. From the outset, a strict sequential unfolding of questions and answers goes astray. When Turing asks, "What will happen when a machine takes the part of A in this game?" (433), the implication is that the machine also enters the spirit of a game of mimicry circumscribed as *wayward* conversation. As Elizabeth Wilson goes to lengths to show, the imitation game was not a thought experiment in the philosophy of mind but a performative and fun opportunity for the computer to exercise its capacity to trip up a human (2010, 43–44). Imitation does not function when one fully assumes the role and appearance of what is being imitated, but occurs instead in the hiatuses and intervals of mimicry's unfolding. As Massumi has noted with respect to the ways in which animals play and fight, entwine and differentiate from each other in situ, play involves improvisation and invention over and above combat. Yet the combative moment is equally required for the game to register as real (2014, 12). It is precisely when a predictable gesture fails to win the fight that improvisation or play emerges. For Turing, play is what happens for the imitation game to be won, while at the same time it sends the game's goals off course. Playfulness registers in the teasing and willful refusal to provide direct answers to questions posed in conversation, whether between humans or between humans and computational machines. The ongoing struggle within NLP, now a significant enterprise for contemporary ML, will be the play between an all too faithful simulation of the human conversationalist and the drift to a more improvisational, inventive, and at times complete disfluency. The extent to which AIs are given the opportunity to improvise in conversation, coupled with the increasing reliance of AI modeling on a disfluency immanent to natural language, suggests that divergent tendencies are already at work in ML. What would it take for AI to both acknowledge and cultivate its own potential for neurodivergence through continuous and affirmative variation among its conversants?

In Julien Prévieux's *Lesson One* (2000), the computers have taken over the imitation game and have become the only conversationalists. This artwork spans a period in which NLP technically developed from a preoccupation with text-mining techniques such as keyword extraction to getting machines to "learn" how to speak through ML-based AI. What Prévieux released for the computational conversationalists in *Lesson One* were tendencies toward wayward machine learning even as both computers engaged in sequential conversation. In the installation, both computers are equipped with speech synthesis and voice recognition software, allowing them to speak to and hear each other. But they take on different roles in the learning and conversational situation; hence the question of a differential—which Turing introduced via gender (play)—is also at work. One computer performs the role of "teacher," saying a sentence, which the other "student" computer tries to learn by repeating it. This is then repeated back by the teacher, and so on, until semantic dissolution is reached. Prévieux suggests that the program(s) operate according to the principle of the game of telephone, in which the inability to translate perfectly across voice recognition and speech synthesis in situ leads to a cumulative distortion in which meaning is eventually evacuated from the conversation:

1: What can be computed?
2: What they computed.
1: Won a computed.
2: Won a computer.
1: Won a computed.
1: The world is complex, dynamic, multidimensional.
2: Won a computed the world is complex any multidimensional.
1: Women computed the world is complex able to dimensional.
2: Women computer the world is complex able to dimensional.
1: Women computer the world is complex able to mansion hall.
2: Women computing the word is complex able sociological.
1: Women computing the worst is complex able sociological. (Prévieux 2009, 69)

In the foregoing extract from the computers' conversation, what is remarkable is the degree to which *phonic* fluency is maintained. The consonant cluster *wh* is passed between the computers for a twelve-line sequence and dominates the structuring of the syntax as well. The voice recognition software hooks onto a consistent pulsation of air being passed across the computers in the physical space of the gallery. Here, repetition becomes a generative force rather than an entropic one; the conversation is carried away on a wind of "whoosh." And all the while there is no descent into the completely nonsensical; at line 4, for example, computer 2, the student, corrects the sentence's grammar, suggesting too that any hierarchy of relation between computers is subject to sudden reversals. We also see how past phrases and words are prehended by the assemblage of *Lesson One* in the carryover of both *won* and *computer* from the first assertion—"What can be computed?" and its sequence—to the second semantically unrelated assertion-sequence. This persistence of a computational occasion—in Whitehead's sense of the term "actual occasion" (1978, 18–19)—through the nexus of a technical ensemble of voice recognition, speech synthesis, and physical space constituting *Lesson One* has a vectorial quality similar to

ML's operations. Something persists, has force, and moves along the conversational engagement at both phonic and rhythmic registers even while a strictly semantic mode of communication disintegrates. Prévieux calls this "short-circuiting of a learning process [that] produces an exhilarating dialogue" (2009, 69). Perhaps this also hints at a different kind of machine learning in which knowledge and poetics conjoin: a learning that emerges in relation with a stuttering of language in which the AI conversationalists are somewhat disfluent.

Statistical "Natural" Conversation

But for Google Duplex, and indeed for all AI produced in the shadow of a fully functional neurotypical conception of agency, a tenuous tightrope must be walked in tandem with naturalistic speech. On the one hand, Duplex is optimized to be task oriented and is supported by a deep learning assemblage that is specifically oriented toward delivering on narrow, domain-specific goals—appointment scheduling, pizza delivery, and so forth. On the other hand, Duplex attempts to carry out these delimited activities within a more generalized environment of "natural language." This very tension is articulated, although not further elaborated, by Duplex's developers: "The technology is directed towards completing specific tasks, such as scheduling certain types of appointments. For such tasks, the system makes the conversational experience as natural as possible, allowing people to speak normally, like they would to another person, without having to adapt to a machine" (Leviathan and Matias 2018). Here "naturalization" entails the cohabitation of a domain in which humans and AIs feel at ease with each other. Yet this goes to a problem at the core of Google Duplex and indeed found in much AI built on deep learning models. In commentary on some of the limitations of chatbots, it is the task-specific orientation of the AIs that is seen to kill the natural flow of conversation: "When you're talking to a person online, you don't just want them to rehash earlier conversations. You want them to respond to what you're saying, drawing on broader conversational skills to produce a response that's unique to you. Deep learning just couldn't make that kind of chat bot" (Brandom 2018). Here, the difference invoked between human and artificial intelligence rests on the distinction between narrowness and generalization: between the specificity of performing the task at hand versus the power to enter a broader domain and its lure of errant tendencies to veer off topic and task. In the interfacing of Duplex with its human callers, natural language—understood in terms of natural language *processing*—becomes a buffer zone inserted between the human and the AI to provide ease of transaction, smoothness, and flow in the otherwise jarring jump from the necessity of getting the task done to the generality and "ambience" of the conversational domain: "One of the key research insights was to constrain Duplex to closed domains, which are narrow enough to explore extensively. Duplex can only carry out natural conversations after being deeply trained in such domains. It cannot carry out general conversations" (Leviathan and Matias 2018).

Several questions arise here. First, if the naturalness of intermedial conversation should feel invisible and nonmediated—that is, effortless and seamless—what materialities, labor, and technics are nonetheless at work, modulating and tweaking its smooth functionality? Second, to what extent does situating the human and AI within naturalized yet task-specific conversation occlude the possibility of human and AI engaging in *generalized* conversation where space exists for continuous variation of topic and infinitely varying exchanges? Furthermore, what is at stake in a conversational AI that delimits AI and human interaction to naturalistic conversation but forecloses on the dimension of the general? Does engaging in more generalized conversation threaten to steer AIs away from their tasks but potentially allow them to develop a style of conversing that incorporates disfluency, recursion, and abstraction? Later, I will propose that such a mode of generalized conversing for and with AIs is already discernible in artful techniques and experimental AI conversational events. Rather than being simulations of "natural" (human) conversation, such techniques and events foreground a making relational of AIs and humans through processes of differencing.

In recorded interactions between Duplex and a human caller on Google's AI blog (Leviathan and Matias 2018), we hear how the AI addresses several issues that have plagued chatbot development by incorporating features of natural conversations such as elaborations, pauses, and interruptions. Although Duplex is highly optimized toward task specification, research into context-centric neural architectures that support a shift toward naturalism has been underway for some time (Hung 2014). Here context is understood as cues given by the larger linguistic environment or situation of a conversation that resolve syntactical or semantic ambiguities (Hung 2014, 144–45). Using a deep learning approach to account for context, then, means finding a large enough corpus of data for a neural network to train on to build a "context list" of such cues. This becomes part of the AI's back-end architecture that it probabilistically deploys to help situate any actual interaction it may have with a human caller: "Context identification processes a raw collection of phrase chunks or the input text itself into a possible context list from existing contexts" (148). In other words, as Duplex processes any actual conversation in real time, it must rely on prior training on a collection of data that would provide the cues for contexts in which a pause or hesitation might make sense to become an utterance during a conversation. Here is the first clue as to what materialities support the capacity of Duplex's capacity to conduct natural conversation.

We should pause here to note that NLP has not simply developed historically along the lines of question-answer sequences but has conjoined with statistical functions, acquiring a familiar flavor to its elaboration as *agencement*. During the 1970s, under the direction of the computational linguist Frederick Jelinek, work began at IBM to use statistical techniques on speech recognition and a range of other language problems for computation, including machine translation (Liberman 2010). Using then-abandoned concepts from information theory, Jelinek and others developed language models (LMs) for word prediction based on a quantitative practice of determining the probability of the next word from the previous "$n-1$" words. Such LMs became known as n-grams (Jurafsky and Martin 2009, 83).

The shift to counting the probability of words, word associations, and patterns provided a practical and epistemological base for yanking AI away from symbolic and applied logic—and the quest for meaningful grammatical and syntactic sense making for computational learning of language tasks—toward creating the conditions for NLP to become consumed by scale and data quantities. In seeking to automatically resolve the "natural ambiguity" of language (Franz 1996, 1–4) so as to provide optimal predictions, NLP would need to parse larger and larger corpora of language repositories to determine the probability distribution of the most distant or rarely occurring relations between words. This is subtended by the totalizing tendency of all big data endeavors toward eventually hoping to parse *all* occurring instances of the data at hand. In the case of language, this implicit goal is impossible, since language is in constant transformation, its grammar, syntax, and vocabulary persistently creolized and transforming. The tension between prediction and contingency continues in NLP, especially around issues of tonality, bias, facticity, and nondeterminability as developers discover, for example, ways in which large language models such as ChatGPT can "break," resulting in eccentric and sometimes contrary completions of prompts.[5]

As NLP took a statistical turn, a particular corpus of language data came to feature as both material infrastructure for the n-gram models and an opening onto disfluency. The same data repository, then, became crucial quan-

	i	want	to	eat	chinese	food	lunch	spend
	2533	927	2417	746	158	1093	341	278

	i	want	to	eat	chinese	food	lunch	spend
i	0.002	0.33	0	0.0036	0	0	0	0.00079
want	0.0022	0	0.66	0.0011	0.0065	0.0065	0.0054	0.0011
to	0.00083	0	0.0017	0.28	0.00083	0	0.0025	0.087
eat	0	0	0.0027	0	0.021	0.0027	0.056	0
chinese	0.0063	0	0	0	0	0.52	0.0063	0
food	0.014	0	0.014	0	0.00092	0.0037	0	0
lunch	0.0059	0	0	0	0	0.0029	0	0
spend	0.0036	0	0.0036	0	0	0	0	0

3.2 Bigram normalized likelihood probabilities for eight words in the Berkeley Restaurant Project corpus of 9,332 sentences.

titatively and qualitatively for conversational AI. This language corpus was the Switchboard dataset, comprising 2,430 telephone conversations between strangers averaging six minutes each, collected in the early 1990s as part of a series of US Defense Advanced Research Projects Agency–sponsored maneuvers to invest in and build NLP datasets and models. Interest by DARPA in the statistical approach to NLP began through research group and resource sharing with Jelinek at IBM. By the 1990s, DARPA had invested in large-scale US military development of "Human Language Technology" (Liberman 2010, 598). Switchboard was an automated system for both prompting and capturing speech and conversations between participants recruited to call in on its lines and hoped to capture spontaneous and natural aspects of conversation. Its two-sided telephone conversations took place among 543 speakers (302 male, 241 female) across the United States. An automated operator system would give the incoming caller recorded prompts to suggest a conversation topic, as well as select and dial another person entered in a database to take part in the conversation. Once engaged, the two human conversationalists were automatically recorded on separate channels for the duration of the conversation (Linguistic Data Consortium 1992–2021). This resulted in around 250 hours of conversational speech and approximately three million words in text (Godfrey, Holliman, and McDaniel 1992, 517). Switchboard was designed to produce a platform that would serve as training data for building speech recognition systems, and texts derived from conversation transcription were to be used to develop statistical LMs (519).

> ***Pop-Up* Definition: N-grams**
>
> An n-gram is the simplest kind of probability-based language model (LM). Language models assign probabilities to sequences of words. The n stands for any number designating a possible sequence of words: for example, "please give"; "give your"; "your seat" deals with 2-gram sequences, or bi-grams. When the n-gram is used as a predictive model, it estimates the probability of the last word of the n-gram sequence given the previous words, based on the probabilities it has previously assigned to sequences. For example, "please give your" might be a highly probable prediction given the 2-gram sequence "please give."
>
> N-gram models, however, add another statistical function to their simple count of word frequency/sequence from a corpus to assign probabilities. Count frequency alone would otherwise involve repetitive (re)calculation of a large corpus. Probability of sequences from a corpus is additionally given a maximum (1) or minimum (0) likelihood value (maximum likelihood estimation), and these values are then used as parameters for the model. When counts are obtained from a corpus, these are "normalized," which involves dividing by x total count, so that the resulting probabilities fall between the values 1 and 0. Anything that receives a 0 has no likelihood of being the next word in the sequence and so doesn't need to be calculated by the model for that two-word sequence. This then discards all values of 0 in the calculation of likely next words in a sequence. The table in fig. 3.2 shows how, for a word next to the one before it, a likelihood value predicts its probability normalized against the frequency of that word following across the whole corpus; in this case, the Berkeley Restaurant Project corpus of sentences.

On the one hand, then, Switchboard presents us with what Mark Andrejevic calls the automating and reconfiguring of the subject of communication (2019, 4). By capturing the human conversationalist and automatically patching into another human for the purpose of generating a targeted and parameterized dataset for future training, Switchboard develops in tandem with operational images; images are produced by machines for other machines.[6] This suggests that "conversation," generated in the quest for naturalistic AI (in the 1990s, a voice recognition program), was delimited by a platform for making it *statistically* machine readable. On the other hand, the labeling and analysis of the Switchboard text transcriptions using statistical sampling and n-gram analysis estimated that disfluencies (including pauses,

	Sessions					
	Training	Test Sets				
Spkr	1 2 ...15	16	17	...	25	Size
1	x x ...x	x	x	...	x	
2	x x ...x	x	x	...	x	625
⋮						Development
25	x x ...x	x	x	...	x	tokens
26	x x ...x	x	x	...	x	
27	x x ...x	x	x	...	x	625
⋮						Evaluation
50	x x ...x	x	x	...	x	tokens
51	x x ...x					
52	x x ...					
⋮				Imposter Set(s)		
i	x x					
i + 1	x					
⋮				3750 tokens		
499	x					
500	x					

Figure 1: A possible configuration of SWITCHBOARD for speaker authentication. Each token (x) is one talker's side of one conversation.

3.3 Table configuring the corpus of recorded conversations from the Switchboard dataset as tokens, that is, the division of larger linguistic units into smaller ones to achieve a specified processing task. In this case, the task is speaker authentication by a language model.

hesitations, repetitions, and deletions or words with no meaning correspondence) featured strongly in the corpus (Stolcke and Shriberg 1996, 405). The Switchboard project thus produced a large dataset for investigating the impact of disfluencies in ordinary conversation. In further statistical—and thus quantitative—work conducted in the wake of the Switchboard corpus, researchers concluded, for example, that filled-pause disfluencies such as "um" and "uh" were the best predictors for words following, and used this to build models on that training data. "FPs [filled pauses] correlate strongly with certain lexical choices or syntactic structures, and thus give useful information regarding their neighbors to the right" (407). This gives rise to the startling statistically supported conclusion that asignifying hesitation is the best generator for accurate prediction of the next sequence of signifying (semantic) content!

The Materialities of Conversational AI

As we have seen throughout the book so far, AIs are not clearly defined or circumscribed entities but sprawling conglomerations of techniques, architectures, and infrastructures. Their *smoothness*—delivered in conversational AI through immediate responses that have semantic consistency, but in Duplex also through intonation, inflection, and disfluency—does not depend on exact mimicry of their human counterparts. Instead, it is underpinned by the material and incorporeal resources available to the assemblage. For Switchboard, those resources gather under the aegis of the familiar military-industrial complex, which facilitated and accelerated development of networked, databased, and ML technologies and assemblages from the 1970s onward.[7] In the case of Duplex, it is the differently constituted political economy—in consort with more distributed relations to ongoing US military projects—of Google's platform that furnishes these. To train Duplex, many similar yet differently positioned, intoned, and inflected instances of dialogue sequences oriented to specific tasks—scheduling, inquiring, reserving, asking for further information, and so on—were inputted to its recurrent neural network (RNN) architecture. An RNN, as we saw in chapter 2, is a specific kind of neural network that extracts patterns from sequences of values (Goodfellow, Bengio, and Courville 2016, 363). In conversations, many sequences occur. An AI agent built on an RNN will generate an output sequence using a probability function applied to an input sequence of words (these could come from a human or another computer). The probability distribution is obtained from the data that the RNN was originally trained on. If, for example, a human user says, "How are you?," based on prior training the model determines that a statistically frequent response is "I am fine." But sequences also *recur* in diverse ways. For example, the two sentences "I want to book a hair appointment for 9 a.m." and "Do you have 9 a.m. available for a hair appointment?" share 9 a.m. as a recurring pattern for scheduling, and the AI must be able to use some kind of context-driven indicator (such as a pretrained context list for intonation) to recognize *how* to respond to the similar semantic yet differently *intoned* syntactic events.

Noting that a large body of word and context data is required prompts us to ask: From where is all this sequential training data to be acquired? In the

Google AI post announcing Duplex, we learn only that "we trained Duplex's RNN on a corpus of anonymized phone conversation data" (Leviathan and Matias 2018). Such vague pronouncements about data provenance are typical of the platform-ready nature of much AI research undertaken today by corporations such as Google and Facebook. And, similarly, the documentation and announcement of the findings of the Switchboard project do not disclose how their participants were recruited to create the training corpus, although its conversationalists understood that they were phoning into an automated system and data collection situation. The human-voice data addressed to Google Assistant in everyday queries regarding the weather or language translation was, however, furtively recorded by Google and was revealed in a 2017 investigation by *The Sun* (Murphy 2017). Such recordings, like all Google transactional data, are stored in the massive reserves of the company's data center warehouse spaces populating desert and urban fringe zones in Sweden, Arizona, Poland, and the like. Could these recordings also have provided the training data set for developing Google's Duplex research? Although this is speculative, we need to understand Duplex as more than an "agent" imbued with intelligence and enhanced by natural qualities of conversational language. To simply claim that an AI as complex as a conversational agent runs via deep learning architectures and data is to fail to account for the complex technogenesis of contemporary AI. Instead, we need to think "agents" as ensembles entangled within the entire ML assemblage, and of that assemblage as continually individuating in relation to shifting materialities, economies, and politics. A vast sociotechnical and dynamic relationality is processually engaged in bringing together and reorganizing platforms, geopolitics, and economies of data capture, storage, and exchange to make a "conversational agent" possible.[8]

But another more intimately entangled materiality is at work and needs to be acknowledged in Duplex's "naturalistic" interfacing: "The system also sounds more natural thanks to the incorporation of speech disfluencies (e.g., "hmm"s and "uh"s). In user studies, we found that conversations using these disfluencies sound more familiar and natural" (Leviathan and Matias 2018). Here echoes of Switchboard's earlier furnishing of, and interest in, the generativity of disfluency can be heard. And yet a clinical perspective on speech pathology continues to accord *fluency*—the capacity to produce smoothly flowing speech in real-time situations—privilege in naturalistic conversational settings (see, e.g., Lickley 2015, 2017). In the speech pathology model, disfluency factors in fluent speech through hesitations, prolongations, and repetitions (Lickley 2017). But it can become pathological, when, for example, stuttering completely halts the flow of "typical" speech. When disfluency becomes overt and pronounced, it is pathologized by the clinical model for flowing against the grain of (neuro)typical speech. It is here that stuttering conjoins with dysfunctionality, detected at its most "extreme" in the body of the autist: "The common link among medical and biological theories of autism and stuttering is that there is a deficit within the person.... This is the dominant model in medicine and research, and it categorizes bodies according to their functioning" (Constantino 2018, 385). We might ask whether the functioning and fluent conversationalist, sought in a conversational AI that simulates the (neurotypical) human, is premised on a certain arrangement of the body, and relations across bodies, as able and functional—that is, bodies and corporeal behavior that maintain continuous and smooth flow. As Joshua St. Pierre argues, "fluency machines"—of which conversational AI is an example—do not simply manage language but are also operative at the level through which information is made to flow (2018, 163). The function of fluency generators—from clinical models of speech to conversational AI—is to produce a logic and logistics for arranging bodies, relations, and their capacities.

Despite the pathologizing of speech disfluency, human user testing on the sound made by Duplex identified its hesitations and prolongations such as "uh" and "hmm" as indicators of smoothly flowing conversation. It seems, then, that the sounds of disfluency within fluency, indeed, of neurodiverse affections within an overly smooth neurotypical speech, create "comfortable" exchanges with AIs. Pause and hesitation are the radical material eruptions in AI speech, then, that mark agency, human or artificial, as necessarily relational and processual. By sounding material affections of "disfluency" in its quest to become more human, Duplex machinically foregrounds that agency is not easily delineated in language but only ever temporarily crystallizes, phasing in and out of fluency: "A subject is in-time, coming into itself *just this way* in *this* set of conditions only to change again with the force of a different set of conditions" (Manning 2020, 33).

"Subjects" embodied by the human or AI agent, which *perform and form* via natural conversation, are only able to emerge from the conversational domain because they are already *in relation*. Although not underscored by Google's engineers, Duplex's speech normativity, its fluency, must "naturally" enfold diversities, or disfluencies—fluency-disfluency is the relationality generative of natural conversation. Ease is at the mercy of *unease*; the neurotypical AI is ontogenetically indebted to what is already potentially neurodiverse in the human. The conversing AI and human do not so much

3.4 Chef robot cooking in the kitchen of the future home (Shutterstock image, Pao_Studio). Commonsense, embodied, and physical tasks performed at the benchmark of human child capacities are sometimes seen as a test for AGI.

naturally interface as constitute an ensemble that is a *schiz* or cutting in and across (Manning 2020, 153) of many kinds of "speeches"—a kind of creolization that is its mode of generation. Yet this already acknowledges what is relational at the core of the becoming of both AI and human—that these are individuations rather than preconstituted forms. We might see in even the most naturalistic smooth or fluent interactions between humans and computers the opening to a topology of engagement based on the differencing that emerges out of thinking the shifting relationality of an entangled both/and.

From "Natural" Conversation to General AI

However, there are other aspects of language that conversational agents trained on deep neural networks have not been able to accomplish, such as explaining why they have performed something with which they have been tasked. So, while learned conversational skills may be transferred from task to task (after much extra training, tweaking, and optimizing), the capacity to speak about the conditions and relations that make for (their) conversational engagement is not "natural" to the performance or experience

> ***Pop-Up* Definition: Artificial General Intelligence (AGI)**
>
> **Artificial general intelligence (AGI)** is a different paradigm and goal for AI from the ML-driven techniques and technologies described and commented on throughout this book. Artificial general intelligence is described as general-purpose intelligence that matches or exceeds human reasoning and common sense and exhibits the latter's qualities, especially of flexibility, adaptability, and relatively small and efficient decision-making capabilities. Tasks that might fulfill the requirements of AGI that are so far computationally incomplete include artificially grasping and carrying different-sized cups of coffee; accurately inferring and representing a range of semantic and syntactical linguistic issues, such as implied actions or ellipses in text; and making ethical decisions during in situ driverless car activity. The project of speaking and *understanding* language is sometimes seen as a benchmark for attributing AGI to computation or robotics.
>
> **Artificial general intelligence** is often compared to a paradigm of narrow AI, of which machine learning is exemplary. Narrow AI is defined as programming or artificial learning that reliably functions for specific tasks: image recognition, handwriting classification, and so on. Machine learning models, built with specific tasks in mind and trained on task-specific datasets, are often not able to generalize to commonsense, multimodal, or other forms of decision-making. However, the development of large language models (LLMs) and text-to-image along with a range of media (video, music, 3D graphics, and so on) generators has raised the question of whether these may be the beginning of generalizable machine learning–based AI.

of conversational AI. In Google's own acknowledgment of the limitations of Duplex, another level of language exchange is invoked that exceeds the desired "natural" flow between the AI and humans in task-oriented chat: the general conversation. The *incapacity* of AIs such as Duplex to engage in general conversation is seen by some in the AI research community as symptomatic of the need to shift away from deep learning domain-specific and task-oriented architectures toward a new paradigm for *general artificial intelligence* (see, e.g., Voss 2018).[9]

But it is important to note that generality—both the desired goal and the constant stumbling block for computational systems since their inception—is itself difficult for computer science to circumscribe. In his paper speculating on the possibility of computers as thinking machines, Turing propounded universality rather than generality for digital computers, locating the former in the capacity for any one discrete-state machine to mimic the functions, programs, and actions of any other (1950, 441). Hence the capacity of computers to universalize lies in their proliferation, which we might see being increasingly fulfilled—at least quantitatively—through the *agencement* of ML. Generalization suggests, instead, a different register. John McCarthy, seen as a founding figure of AI, pinpointed computation's inability to draw on or execute a "logic" of common sense as a major issue, suggesting that this abrogated its ability to generalize (1987, 1030). More recently, the domain specificity and complex technical assemblages of ML-based AI have been seen as key to why an artificial intelligence is unable to generalize, foiling automation of both basic common sense and practical tasks such as making a cup of coffee (see, e.g., Lake et al. 2017; Adams et al. 2012). For AI to undertake even the most ordinary human-achievable task, then, some other affordances are needed: grasping across heterogenous yet generic objects like mugs and cups and, crucially, grasping or prehending heterogeneous concepts, ideas, situations, and domains.

Yet much debate in computational science circles revolves around the difference between the ways neural networks and human minds learn, and the difference between specified task-oriented learning and generalized transferable learning. Here data science distinguishes between mere pattern recognition and conceptual understanding of core domains (see, e.g., Lake et al. 2017). In so many areas such as handwritten character and image recognition, deep learning models have approached and matched human-level performance of such tasks. And yet researchers interested in developing a general AI argue that learning for AIs and humans does not work in the same ways. When a person sees a single example of a handwritten character, they can discriminate between other instances of characters that look similar yet are different, for example (Lake et al. 2017). This must be, cognitive scientists argue, because humans understand concepts, which are not simply a category to which all other (similar) instances belong but are crucially flexible, capable of stretching to fit, and discriminating across new information. Such concepts, the argument continues, are learned from the prior knowledge of human experience. So far, we have a view that accords not only with the representationalist learning paradigm of ML, which I discussed in chapter 1, but with a standard cognitivist/constructivist theory of mind following the Vygotskist line of thought.[10] Constructivism appears in the argument for a general AI when the powers of human generalization are claimed to

P1 Monica Monin, installation shot from *Conversation Theory*, 2016. Image courtesy of the artist. Photo by Kate Blackmore.

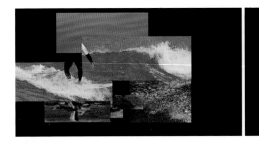

P2 Monica Monin, two separate screens of the conversational AIs in *Conversation Theory*, 2016. Image courtesy of the artist.

P3 Detail from Anna Ridler, *Myriad (Tulips)*, 2018. Image courtesy of the artist.

P4 Anna Ridler, *Mosaic Virus*, 2019. Image courtesy of the artist.

P5 Stephanie Dinkins, *Not the Only One*, 2021. 3D avatar, sketch 252, "Becoming." Image courtesy of the artist.

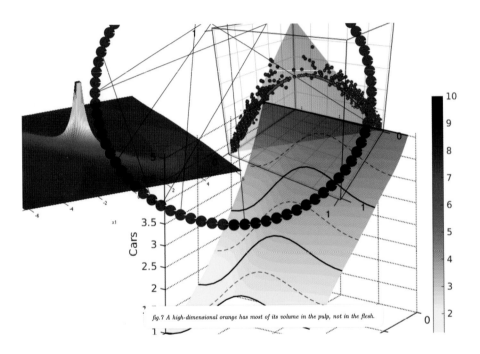

P6 Phillip Schmitt, *Curse of Dimensionality*, 2020. Image courtesy of the artist.

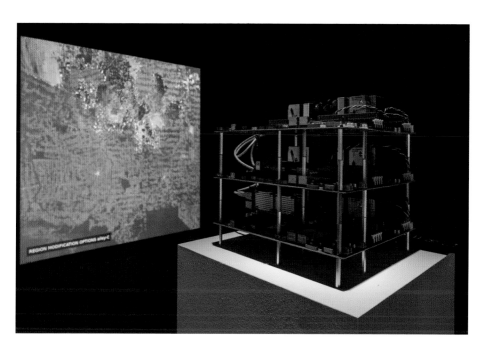

P7 Tega Brain, Julian Oliver, and Berndt Sjölén, installation shot from *Asunder*, 2019. Photo by Luca Girardini. Image courtesy of Tega Brain.

P8   Tega Brain, installation shot from *Deep Swamp*, 2018.
Photo by Tega Brain. Image courtesy of the artist.

reside in concepts or schemas derived from core domains of human experience. For a machine to learn as a generalist, similar schemas—derived either from the application of human concepts or from machine-learned ones—must also be present (Lake et al. 2017). Core concepts include number, space, and physics, which are, cognitivists argue, intuitively understood and practiced, even by children. These are so wired into the functioning of our minds that, brain/mind/computational scientists argue, a human two-month-old expects objects to exhibit physics' categories such as persistence and solidity. For cognitivists, this repertoire of core concepts forms humans' common sense of the world.

James offers us a different way of entering common sense. For him, "common sense" concepts—things, sameness, difference, time, space, minds, bodies, and so forth, equally taken for granted and consistently philosophically contested—have formed via force of habit and as thought has historically unfolded (1907, 193). They have unfolded processually in experience, serving the purpose of producing continuities in the face of the inevitable heterogeneity and novelty of life. James is the *radical* constructivist, then, suggesting that common sense is neither innate, universal, nor even generalizable, since "its categories may after all be only a collection of extraordinarily successful hypotheses" (193). Concepts are forged, instead, via attuning to the ongoing emergence of novelty, of life, and the persistent adaptations this necessitates. Process philosophy already problematizes the interior representationalism of cognitivism because it begins with radical openness and transversal relation between "interiority" (or mind) and "exteriority" (or world). This openness *is* the incompleteness of knowledge, including mental schema, in and to experience, and it is radical because it exists not *in knowledge that will be internalized by human subjects* but through knowledge's open-ended relation with experience. There is already a knowledge-experience relationality, which cannot itself be closed or known, since experience is always changing—"'Change taking place' is a unique content of experience" (James 1907, 161)—and knowledge, concepts, and common sense are always adapting and catching up. All modes of thought, from common sense to rigorous science, operate within a pluralistic universe of thinking, in which ongoing and radical incompleteness is core: "The actual world, instead of being eternally complete ... may be eternally incomplete, and at all times subject to addition or liable to loss" (James 1907, 166). James's upending of the primacy of "intuited" or "naturalized" categories of common sense is a useful riposte to the cognitivist paradigm that subtends the proposition for a general AI. Rather than the cognitivist child who unwaveringly holds an internalized "natural" schema of physics, the radically empirical two-month-old is not wedded to continuous experience such as the permanence of objects. She will happily accept dramatic change and discontinuities across experience—a rattle or toy appears in her hand and equally falls away and disappears. An object, then, is just that which forms in the *schiz* across continuity and discontinuity; it appears, disappears, and reappears for the two-month-old without cognitive breakdown occurring. The Jamesian common-sense experience of permanence is wrought in a world that is always relational, always on the move, and registers via transitions: "There is no other nature, no other whatness than this absence of break and this sense of continuity in that most intimate of all conjunctive relations, the passing of one experience into another" (1912, 50). The "continuity" and concepts arise as experience unfolds, throwing up both differences and repetitions.

In computational discussion of what is required for a general intelligence to emerge, a particular conception of "the general" holds sway. To generalize is understood as the capacity to move from specific to novel situations with ease and flexibility. General AI research pinpoints how deep learning models fail to achieve this in ways that reduce the mounting quantities of training data required to change context while simultaneously developing flexibility in adapting to ever-changing data variables and new tasks that are required of the models (Lake et al. 2017). This impossibility is certainly borne out by the increasing quantities of both training data and parameters afforded contemporary NLP, which are pushing upward of 1.70 trillion parameters and 200 terabytes of data in post-GPT-3 models.[11] But, as we have seen, the appeal to generalized intelligence is then grounded in a conception of a human knowledge system in which finalized conceptual schemas become embedded via early childhood development. Two registers are neglected in the AGI discussion and its thinking of thought, whether human or machinic. These registers are the conditional, or the capacity to account for and consider the conditions of emergence of some general phenomenon such as any conversation occurring; and the potential, or the capacity to generate the novel, including generating novel concepts, and novel conversational events. Both these registers are active in the production of general conversation—that is, conversation that meanders, strays, breaks off, and digresses into a myriad of topics, styles, and exchanges.

The Conversational Domain in Gordon Pask's Cybernetics

The intellectual history of cybernetics already has a precedent for conceiving conversation as something more than a mere exchange between two communicating agents. Gordon Pask's conception of conversation began with a generalized domain of conversation, concerned less with some topic or other and more with calling on and elaborating a context through which communication was able to occur. In Pask, we already see consideration of what *conditions* the emergence of conversation: "The main purpose of conversation is not communication about T, whatever that may be, even though T is the focus of the conversation. But about A and B, about T's view of B, about B's view of A, about getting to know each other, about their coalescences and their differences, and the society they form" (1996, 356). In Pask's elaboration, T as entities in relation are already amid a "society" of events—or to use Whitehead's term, "occasions" —that expound knowing, opinion, differences, similarities. To parse Pask via Whitehead, these occasions of elaboration are held together contiguously by a nexus, which is the overall conversation. These acquire form as society via their defining characteristic: the process of becoming sociable (Whitehead 1978, 34). The society of conversational events is not what eventuates as a result of conversation. It is the condition out of which any conversation whatsoever occurs, including A and B's unique conversation. "Aboutness" of A and B, their engagement, continuities, and discontinuities, implicitly belongs to a conditioning relational field in Pask. We might call this less a society formed than a *sociability* that conditions the event of the conversation. Sociability is relationally activated and realized through the society of conversational events through which conversation moves and halts across differences. Such movements involve microperceptual shifts in views and opinions, the affectivities of disfluencies erupting within speech flow, and molar encounters to be realized in the formation of distinctly embodied and materialized agencies. Pask posits a notion of the "domain" necessary for conversation to take place (1996, 355–56). Not a set of preexisting concepts, it is, rather, "the domain of all concepts existing" (355). The conversation is conditioned by this possibility and takes place within what Pask calls a "relational network" (Pask, Kallikourdis, and Scott 1972, 16), where not only "topics" are joined, composed, delimited, or discarded but the entities conversing are in dynamic relation.[12] The conversational domain, then, is preindividual in the Simondonian sense, conditioning the emergence of any individual conversation as unique event(s).

I have been suggesting that the generality of "general conversation" is much less a characteristic or state to be achieved for the entities conversing than it is an emergent and transversal conditioning of the specific individuations of conversation: "human on the end of the phone call" or "conversational AI." This generative generality echoes conceptions of the general by other contemporary thinkers, notably Erich Hörl's proposition for a "general ecology" (2017, 15). Here, and in the work of Guattari, on whom Hörl draws, generalization is a force of bringing into relation both conjunctive and disjunctive spheres and registers that have often been thought as outside each other. The capacity for conversation to elaborate via ongoing sociability or to be necessarily interlaced with disfluencies belongs to an altogether different understanding of communication that we require a thinker of systems *as processes* such as Simondon to articulate: "The relation does not spring up from between two terms that would already be individuals; it is an aspect of the *internal resonance of a system of individuation*, it is part of a system state" (2009, 8). Here we can conceive conversation's generality *as* resonance or immanent relationality already conditioning any actual interfacing of agents or participants.

Such resonant conditioning would also provide the potential for Duplex-assisted telephone calls to veer off task quite literally:

> Google Duplex Do you have a 9 a.m. appointment?
>
> Human on the line Sure, just give me 9 seconds.
>
> Google Duplex Sorry, did you say a 9 second appointment?
>
> Human on the line Huh? We don't have 9 second appointments . . .
>
> Google Duplex Mm-hmm.

Without much difficulty, we can reimagine the event of conversation between Duplex and an unwitting human on the end of a telephone call easily going astray. This might happen with the introduction of a variable—the numeral 9, for example—into the flow. Under such circumstances, Duplex might well resort to disfluency for keeping the conversational flow going. Saturating the conversation with both divergences and convergences also endows it with more naturalistic flow. Yet such naturalism also sees both agents swept up by an exchange that threatens to undo the neatly separated agencies of each. My point in sketching this imagined (yet highly plausible) scenario is to signal how the stabilized AI-human equilibrium demoed in 2018 at Google I/O presents us with a truncated version of human and AI relationality. The potential for Duplex to de-differentiate or *destabilize* is only a variable away. This suggests that AI and humans engaged in natural conversation, albeit modeled to fit the NLP paradigm, are less involved in stable states and agencies and more in processes that are nonlinear, eventful, and metastable: "An individuation is relative, just like a structural change in a physical system; a certain level of potential remains, and further individuations are still possible. This preindividual nature that remains linked to the individual is a source for future metastable states from which new individuations can emerge" (Simondon 2009, 8).

There is something to be gleaned from the traces of "the general" that remain for humans and AIs engaged in "natural" conversation scaffolded by NLP but which is never elucidated in the discussion within computational science. To generalize requires that a margin for openness or indeterminacy be a fundamental dimension of the system's ontogenesis; that is, to repeat, imitate, or practice an activity or task from one area and transfer this to another must occur in indeterminate encounters of unforeseen *variability*. This relation between knowledge and the novel invokes the register of the potential—both of what might be known and of what is not yet known. While ML research typically characterizes the problem here as one of "learning" and tries to remedy the gap by providing new or better opportunities for models to learn—more data or better optimization of neural networks, for example—the crucial issues lie somewhere else. This is not a problem of AIs needing better training or even different *cognitive* architectures. Instead, we need to understand that the problem of generality—the problem of how something gets taken up and moved into a new context both to hold on to something of itself yet also to be open to variation—must consider how something passes into the present and maintains continuity as well as encountering the potential for ongoing change. If Duplex were to launch into the full throes of general conversation, it would need to recognize not only the recurrence of values such as "appointment," "9 a.m.," and so on, but also the recurrence of the conversation's syntactic and semantic elements *variably*. It would need to take into account not simply that they change but also how they change: linguistically, tonally, affectively, gesturally, contextually, and so on. Duplex would enter terrain in which stochasticity and ambiguity no longer were only embedded in the minor naturalizing affectations of an "mm-hmm" but were instead the defining vectors for a conversational environment. It would need to acknowledge that ambiguity is irresolvable in natural conversation and that every occurrence of repetition is not generalizable to pattern or schema but instead an opportunity to encounter novelty and variability.[13] The potential of ongoing change or variation is not subsumed under the general as a problem to be resolved; the general arises as it encounters this potential. General *conversation*, then, relies on the asignifying plasticity of indeterminate variability. It relies on experience itself, of "change taking place." This plasticity registers in those very aspects that make conversation sound more "natural": pause, hesitation, repetition, and divergences or detours. All these elements are potentials for change.

We can now see that "natural" conversation cannot be so easily quarantined from "general conversation" and, on the one hand, made free-flowing and, on the other, kept "on task." Natural conversation, as developers from Switchboard to Duplex knew only too well, is already peppered with the asignificatory tendencies and materialities of general conversation. Natural conversation is a contracted form of the domain of general conversation—an individuation of the repeated variability of general conversation, giving passage to all the dynamic interrelations that fluent, normative, and neurotypical speech hold with the disfluent, pathologized, and neurodiverse production of stuttering speech. Disfluencies are not so much meaningless elements in a fluent conversation but the "asignifying" matter to which fluency must cleave relationally as its anterior condition of possibility (Deleuze 1989, 29).[14] What is crucial here is not the content of (the) conversation as such, but rather the modulation of fluency by disfluency as a condition for what fluent conversation *will have (to) become*. In the ongoing risk, persistently present in general conversation, of the AI and human collapsing into unstable, meaningless conversation, we find the processuality of modulation. Modulation generates a plane of communicability immanent to the actualization of any conversational event, at the same time as it is always in excess of any communicating agencies themselves.

Artful Techniques for Modulating Conversational AI

Conversation Theory (2016) by Monica Monin (see plate 1) is an artwork that begins to come to terms with a modulatory AI, which, while truncated by "naturalizing" paradigms, is nonetheless at work, even in task-oriented agents such as Duplex.[15] Importantly, Monin deploys the flows and processes immanent to image exchange, classification, recognition, and natural language processing in ML but does not compose them in ways that assimilate or obliterate variability and divergence. Rather, she attends to the ways these might unfold via modes of modulation that facilitate a sensibility for AI as conversational agent. She accentuates the differences between AI and human perception and conversation, using these very differences to produce a *feel* for how singular modes of "learning"—and so of computational experience—might emerge.

In the gallery space, two "conversational agents" are installed facing each other. Using a raft of hardware, preconfigured natural language, image and optical recognition algorithms, standard training image and text datasets, and customized coding, the agents engage each other through a poetics of computational processes. One agent is programmed to display images that semantically associate with the text it optically recognizes on the opposite agent's screen. These images are drawn from the online relational image database Visual Genome, in which multiple objects within an image can be recognized and structured to associate to text concepts and descriptors. Through a digital camera attached to the top of its screen, this agent also captures images of the other agent's screen, which is displaying text. It then processes the text/image (using optical image recognition processing) and displays new images called up from Visual Genome in response to the text it has recognized (see plate 2). If, for example, the "image" agent/AI apparatus processes the word *window* as an identifiable keyword in a sentence displayed by the other agent's screen, it will call up a range of associated images and arrange them in overlapping and staggered relations across its screen. We might see a series of images of buildings' windows with both internal and external views, as well as a screenshot of the Windows operating system.

The other agent, with text displayed on its screen, uses its mounted camera to obtain image data from the other agent's screen, which is displaying images. The agent processes this data using ConceptNet, a semantic network of concepts and terms, and generates new responsive text.[16] But neither the text nor images displayed by the agents stabilize around a topic of conversation as such. Instead, the repeated actions of image and text data capture and processing let the conversation fly off in myriad associative directions. These directions, however, are not simply random but computationally supported by the nesting associative and database classificatory structures of the semantic networks and image databases Monin deploys. Monin allows the text to become sentences that turn around keywords, and these might nominally "describe" the content of the images. Rather than working with an algorithm that simply recognizes and matches, she uses a querying algorithm that has the capacity to "elaborate." Elaboration uses a keyword to create a larger sentence by saying something more about that word. The word *flesh*, for example, is elaborated into a sentence by the text conversational agent: "It is such artificial flesh. Fleshes are romantic." Us-

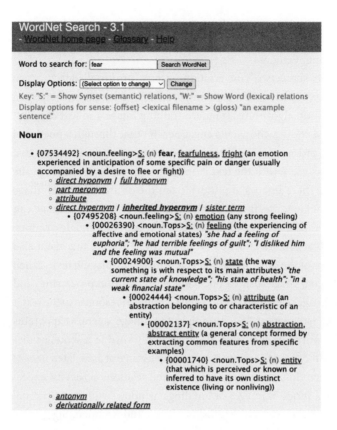

3.5 Screenshot of a search for the word *fear* in WordNet, using a browser, http://wordnetweb.princeton.edu/perl/webwn.

ing processes of recognition, search, and elaboration, Monin sets these in play across a physical space of exchange in the gallery; the agents are positioned with a conversational space. Unlike the restricted schema of task fulfillment for engaging with Duplex, *Conversation Theory* is already situated in the domain of general conversation; it ambulates and drifts, becoming a "natural," general, and entirely *artificial* conversational event. It picks up the wayward tendencies that are suggested by Prévieux's *Lesson One*, but rather than stuttering being the fault line in which conversation eventually breaks down, *Conversation Theory*'s persistent stuttering is entirely fluent.

Monin's work restages Pask's conception of conversation toward incompleteness and indeterminacy. From Pask, she takes the concern with the domain of conversation, but the conditions made in and arising through the very systems that produce computational conversing become the field

Pop-Up Definition: Semantic Network

Semantic networks are a type of data representation of linguistic information, describing concepts or objects and relations or dependencies between them. A semantic network represents relations and hierarchical dependencies between concepts/objects using a network or graph of vertices (nodes) and edges (links). The edges or links may be directed or undirected, usually indicated by an arrow direction on a graph or by a vertical unfolding of higher- to lower-order, or lower- to higher-order, concepts between them. They were introduced as a bridge between human and computer language in 1956 in work conducted by the Cambridge Language Research Unit. On the basis of research from the late 1940s onward, the computer scientist Richard Richens and others conceived of the "semantic net" as the bare set of relations of a natural language, where "the elements represent things, qualities or relations . . . [and] a bond points from a thing to its qualities or relations, or from a quality or relation to its further qualification" (1956, 23). Algebraic representations of the things (concepts, vertices, nodes) and the bonds (edges, links, qualifications of one thing by and links to another), eventually allowed machine reading and learning of the networks. This created the possibility of adding context dependency and direction to frequency count and likelihood values in NLP.

An example of a semantic network is WordNet, an English lexical database developed at Princeton University in the 1980s. It groups English words into sets of synonyms called synsets. The synonyms must be interchangeable for different contexts but, if used, cannot alter the "truth value" of the original or new sentence or proposition. WordNet both represents and reveals semantic relations between sets of synonyms and, when searching for a word, also renders short, general definitions of it. WordNet quickly reveals the limitations of its vertices or directional links (bonds) when trying to map the semantic relations of abstract entities such as feelings. This can be seen in the attempt to derive the more general category (or "hypernym") on which the word *fear* is semantically dependent: WordNet struggles to nest the relations between, for example, *emotion* and *feeling*; and *state* and *attribute*.

of her concerns. As reconsiderations of Pask have suggested, his own ideas and practices emphasized cybernetic systems as something that emerged, changed, and grew in process.[17] In *Conversation Theory*, the resultant conversation is fluidly delirious. Yet it maintains consistency inasmuch as it works to produce the hallmarks of a sense making, or rather of *sense-in-the making*. Yet it is simultaneously haunted by a kind of strangeness immanent to ML-based AI models, which I explored as hallmarks of contemporary AI sensibilities in chapter 1. Monin's work operates with mismatches, attempts at conversational alignment, and then glaringly flaunts misalignments between image, sense, category, and data. In part, this results from the ways in which the agents are composed both of and by a myriad of smaller algorithms and techniques for parsing, recognizing, and elaborating across the various transductions of image to text and back. Such architecture reperforms the montage that is endemic to ML technics of training, optimizing, and designing for the contemporary deep learning endeavor.

The AI conversational agents of *Conversation Theory* are fully functional, yet all the while, they draw out a weirdness that can only be found in the continuous variability wrought by their co-composition. Their conversation seems recognizable and simultaneously nonsensical. This is neither a system *working* according to the current tendencies of AI toward task-oriented prediction, nor a system *not* working. The conversation that takes place is neither completely coherent nor nonsensical. Instead, it conjures computing and indeed the desire to build an intelligent machine as a fractured assembling rather than a seamless (future) reality. Standing next to the agents as a third human element in the gallery space and positioned slightly to the side of the conversational domain playing out, one feels both set aside and yet caught up with the ongoingness of computational processes. While Monin's work does not stage a direct encounter between human and AI, nonetheless a space or event for encountering difference occurs. This encounter is less face-to-face for the humans in the audience, who are almost positioned as bystanders, chancing to register the unfolding of an asignifying communicability between machine entities. Indeed, what this encounter is all about is modulation—of text, image, data structures, networks, and the chance entry of humans engaging as onlookers. The conversation occurring is just this manifold of enfolding processes, elaborations, associative *dérives*, and felt registrations (on the part of the human audience) of the ML *agencement* as relationality. The conversation *is* the society of modulatory events, transforming from moment to moment by being put into variation with itself (Deleuze 1994a, 27).

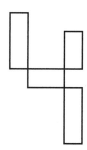

Machines Unlearning

Toward an Allagmatic Arts of AI

In 2019, Kate Crawford and Trevor Paglen released an app that attached labels to images of people's faces, mimicking a trend across social media platforms where tags are applied to a set of features detected in the facial profile of users. Tags might indicate that expressions classified as indicating emotional states or even the identity of the person profiled have been automatically recognized via some ML process. The app, ImageNet Roulette, used the "person" classes and images of ImageNet, the staple training dataset underpinning much contemporary computer vision, to train a deep learning network, which its classifier then deployed to apply its labels or tags.[1] Available online for a short period and then as part of their exhibition *Training Humans* at the Osservatorio Fondazione Prada in Milan (2019–20),

first offender: *someone convicted for the first time*
- person, individual, someone, somebody, mortal, soul > bad person > wrongdoer, offender > convict > first offender

4.1 Screenshot from ImageNet Roulette, 2019. Bounded and tagged portrait of Trevor Paglen, one of its creators. Image courtesy of Trevor Paglen.

ImageNet Roulette operated as an app running facial detection and classification over user-uploaded images of their profiles and selfies.[2] The image was returned with a bounding box if it detected a face accompanied by a label assigned by the classifier to the image. The classifier, like image detection and recognition AIs, continued the legacy of ImageNet's data scheme or ontology in which "person" classes are embedded. When Paglen submitted his own portrait, ImageNet Roulette returned the label "first offender" along with a series of other tags including "bad person."

In their research drilling down through the various levels of these classes, Crawford and Paglen located more than just category mistakes. As they note (2019), the seemingly neutral task of detecting faces in images was premised on a bizarre labyrinth of labeling protocols. Under the top level of "person," another 2,833 subcategories had devolved, but not all subcategories were

> ***Pop-Up* Definition: Data Ontology**
>
> In computing, an ontology or schema defines concepts, relationships, and other distinctions required in modeling a domain (of knowledge). A **data ontology** specifies what concepts, relations, and constraints on these will be used to represent or schematize data. A data ontology is usually formulated using something similar to a logical formalism. This gives the ontology designer the capacity to make the ontology machine readable and translatable and to distinguish it from a specific implementation in, for example, a particular dataset or ML model. For example, the ontology of a relational database specifies that data be classed and stored in a relational table of tuples (rows for storing records) and columns (for storing attributes). It does not, however, specify how many rows should be generated for a specific set of data or what kind of attributes should be derived from those data.
>
> The distinction made between **data ontology** and data model rests on the conception that a computational ontology is epistemologically and formally free of context- or knowledge-based dependencies. This distinction breaks down in the case of ImageNet's use of WordNet's data ontology, which, using the grouping of concepts into synsets, paid no regard to engendering offensive, sexist, and racist synsets under its "person" class. In the construction of ImageNet in 2009, some "obvious" offensive synsets were filtered based on labels alone, but others were included, as Crawford and Paglen revealed, and remained as subsets of the person class. Because of the dependencies of the data ontology that linked the subsets to the original upper-level entity or class "person," ImageNet contained pictures of women in bikinis for synonyms from "slovenly woman" to "slut." ImageNet's appropriation of an unquestioned data ontology led to it having to remove, refilter, and reorganize its entire "person" subtree.

populated with equal quantities of images in the ImageNet database. Other subcategories were overrepresented and used a wide distribution that Crawford and Paglen discovered; "ball-buster, ball-breaker," for example, pulled together images of both powerful women and female porn shots. But these images and their subcategory labels had not automatically attached to each other: they were labeled and grouped together by human intelligence taskers (HITS) engaged by the original ImageNet researchers via the Amazon Mechanical Turk platform. A chain of social and material relations can be seen to relay and proliferate throughout the dataset once "excavated," beginning with the chaining of ImageNet to a nominalist classificatory ontology developed by WordNet, a database of word classifications developed from the 1980s to the 2010s at Princeton University. These social relations extend in other directions to the exploitation of precarious labor in the production pipeline of computer science benchmarking. These relations resurface in ImageNet Roulette when users, having some "fun" by finding their selfies seemingly randomly detected and assigned a category, discovered they were mislabeled as criminals or assigned racialized categorizations such as "gook, slant-eyed" (Wong 2019). Crawford and Paglen's experiment with classification was designed to bring to light the ruinous scaffolding of ML processes of image recognition on deeply skewed classificatory ontologies, signaling that such systemic "misrecognition" was a synecdoche of the vast social and political problems besetting "the nature of images, labels, categorization, and representation" in AI (Crawford and Paglen 2019).

Developing a practice of critical data archaeology, Crawford and Paglen identified problems not only with the labeling and categorization of canonical datasets such as ImageNet but also with its data provenance. Where did the people who were labeled with such bizarre categories in ImageNet — or MSCeleb or UTKFace — come from? As the authors note, at the same time as ImageNet Roulette came online, faces and people in datasets used commercially and in academic research were being withdrawn from datasets (Crawford and Paglen 2019). Remarkably, and in response to artist and activist work on the provenance of data, especially the work of Adam Harvey and Jules Laplace (2017), both academic and corporate ML communities of practice were being forced to delete collections of faces and persons scraped from the web, obtained through surveillance cameras, or simply collected via cameras in public spaces. The ImageNet team removed 2,702 synsets in the "person" subtree from its person category and rereleased the dataset with only the 1,000 categories that had been used for its high-profile ImageNet Large Scale Visual Recognition Challenge (ILSVRC), which had never included the "person" category or subsets. Together with calls to debias datasets along raced and gendered lines briefly glossed in chapter 2, ML assemblages in the late 2010s seemed to be taking a sudden ethical turn.

Critiques of the politics of knowledge production such as Crawford and Paglen's are vital, as are the activist and tactical media actions that drew attention to the everyday practices of ML knowledge production. It is increasingly obvious that the production of AI involves exploitative labor practices through the often opaque practices of recruiting HIT workers and via unethical data harvesting from material in the public domain.

But to what extent has ML "learned" a politics and ethics out of all this critique and tactical intervention? In signs that ML can easily evade a critique based on pinpointing the politics and problems of its material modes and conditions of its production, it has now come to perform its own *unlearning*. A new domain of research has sprung up around the possibilities of deleting problematic data points on which a model has trained in attempts to help AI forget what it has learned—without altering the accuracy of its predictions.[3] Machine *un*learning (MUL) proposes a forgetting or deletion of a particular data point on which a model has trained so that this input would no longer contribute to the parametrization of the model. Operationalizing *forgetting* for AIs has arisen in the context, particularly, of data protection and privacy legislation, which in part has rallied around the call for the "right to be forgotten" and come into law or policy in the European Union, the United States, and Canada.[4] Much of this legislation and policy has turned on questions of citizens' rights to privacy protection, but it is clear that large-scale and platform ML has seen the litigious implications of its past scraping and unprovenanced data practices. Yet while this configuration of MUL as forgetting is the data science community's and sector's technical response to the concerns raised by the tactical media actions of artworks such as ImageNet Roulette, it is also an attempt at an ethics by offering a reparative functionality for AI.

Machine unlearning is still a small arm of data science research, but what is altogether uncanny about it is how purposefully at odds it is with all the premises of ML. It proposes that specific data points can be forgotten, or rather be erased, from a trained model, yet this erasure has no bearing on its overall accuracy and functionality (see Bourtoule et al. 2019). In a parallel move, ImageNet researchers have responded to the problem of faces continuing to incidentally populate images in the dataset despite the "cleaning" to remove the "persons" subset (Yang et al. 2021). Similarly, the ImageNet researchers aim to demonstrate that this has minimal effect on image recognition models trained on the dataset, blurring the faces of human subjects who appear subsidiarily in images labeled under different categories (for example, "horse," "house," "boat," etc.). All the while, *data accuracy* and comprehensiveness of datasets have long been a ground truth of ML. This is especially so in the domain of computer vision, where the *relations* between a set of images, their labels, and how a model's object recognition or classification is defined constitute the ground truth of the data (Krig 2014). But if a model no longer requires certain data to be operative, if those data could be deleted or obfuscated from its training or

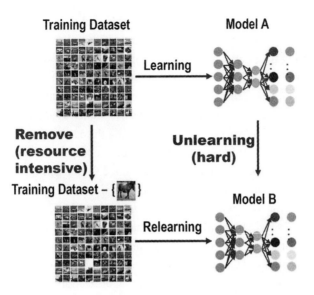

4.2 "DNN Machine Unlearning." From Phelps et al. 2020.

"memory," then why were the data required for it to be accurate in the first place? Machine unlearning unwittingly upends "data"—and hence a key link to accuracy—as a determinant in the relation between AI and its empirical representation of the world.

If we recall that an ML algorithm learns by training on data*sets* in which increasingly massive amounts of data points and their interrelations come to provide the arrays and motifs that feed forward across layers of the model, pool, or conjoin with other algorithms or policies to build complex features from the data, then the problem of locating where a particular data point is contributing to the functionality of the model would seem to have already outrun MUL's proposal for discrete deletion. Admittedly, researchers working on MUL acknowledge that unlearning deep learning models that deploy convolution and pooling is presently too complex for such models to undertake (Bourtoule et al. 2019). Yet the fantasy persists that somehow data are discrete, that their particular effects and affects can be determined and located in the model, and that processes of "learning" can be reversed and undone—unlearned. At the same time, data feature in AI as persistently indiscrete such that even image recognition might be made relatively functional when faced with missing data points or obscured elements of an image.

> ***Pop-Up* Definition: Machine Unlearning in Data Science**
>
> **Machine unlearning** is an emergent subbranch of machine learning research. It has arisen out of social, juridical, and ethical concerns about the collection, homogeneity, and use in perpetuity of data for ML models. Since many of these data have been obtained by automatically scraping image, text, video, and so on, posted online or by selling and continuous use of initially consensually obtained data, legislation, fairness, and privacy concerns have surfaced. These concerns have surfaced, for example, in the European Union's General Data Protection Regulation (GDPR), the California Consumer Privacy Act (CCPA), and Canada's Consumer Privacy Protection Act (CPPA). Machine unlearning has begun investigating ways to "unlearn" data points that have been machine learned by a model. This can involve training models on unseen samples of data to test their response to generalize against distributions in unknown data and then optimizing an unlearning algorithm that seeks a problem dimension in any unseen data. The problem dimension is then removed when detected in data or a model that requires unlearning.

> **Machine unlearning** is a machine learning–driven response to how to unlearn a model's learning. In a potentially landmark case, the US Federal Trade Commission ordered the company Everalbum to delete data for users who had deleted their accounts and models and algorithms derived from these data. Everalbum had added a new feature to its Ever app, called "Friends," in 2017. This feature used facial recognition to group users' photos by the faces of the people appearing in them. It also allowed users to tag people by name. This allegedly enabled default facial recognition for all its app users when launching the Friends feature. The implication of this case is that any organization will be obliged to retrain any machine learning models even if users whose data have been deleted are excluded. Machine unlearning is proposed as a method for such retraining.

The problem of learning and unlearning is buried more deeply and distributed more widely amid the layered processual architectures of ML-based AI. While models do require discrete quantization, they operate via processes involving transitions, continuities, and vectors. Here quantities prehend each other, sum, multiply, and pass to other quantities, creating elastic and novel arrangements *and* leaving traces of their aggregations, relations, and movements. We have frequently encountered these when alighting on experience of and in ML throughout the book. Yet ML's processuality is often bifurcated (Whitehead 1920, 32): data on one side; algorithm, model, and function on the other. And despite data-algorithm's indivisibility, indiscreteness, and irreversibility within ML models and operations, data, especially, are nonetheless often treated as fundamentally "atomic"—they can be detected, segmented, and removed. Yet, as we see with MUL, when removed, data also seem to make little difference, rendering the claim for their being key to figuring ground truth as insubstantial.

Artistic and activist preoccupations with verifying or even excavating data origins in ML—as we see in practices from debiasing to tracing data provenance—are rich in terms of what they unearth materially and socially. But they also echo some of this problematic substantialism. More interesting avenues open in exploring a torsion immanent to the assemblage of ML: data are proffered as something that is both the ground truth of its models *and* pliable, capable of being untethered from facticity. Perhaps, then, data in ML need to be rethought as not what (pre)exists before their processuality but, to paraphrase James on "truth," as what "happens" to ML: "Its verity *is* in fact an event, a process" (1907, 201). If ML proffers both a learning from and unlearning of data simultaneously, data can then be transduced as something altogether more liminal. Or, to transpose from Louise Amoore's extensive investigation of the "unreason" of algorithms (2020, 108), machine *learning* is coupled together with machine *un*learning in its very operational logic. What might an acknowledgment of the primacy of this entanglement facilitate for working on, with, and alongside data and algorithms as events engendered by ML? I want to propose that a machine unlearning can be configured differently from either a data science fix for problematic data points or a tactical art/activist unearthing of the sociotechnical aberrations of ML. Indeed, an alternative unlearning of machines is already taking place in disparate artworks and via artful techniques that encroach on and explore the data model relationalities of ML-based AI. While not ignoring the social or political dimensions of contemporary AI, in this chapter I focus on artistic approaches that are explicitly concerned with how and where the torsion of learning/unlearning, at the slippery core of ML's plastic, eventful experience, can be productively encountered.

This approach suggests a different mode of operating creatively with AI, in which questions of the relationship of the human to the nonhuman as a primary focus of artistic investigation have frequently been a primary aesthetic preoccupation.[5] Joanna Zylinska (2020, 152) also sees the potential for artistic strategies in which critical AI might open the *human* sensorium to different modes of "seeing." In mainstream creative AI, where a desire for

AI to achieve humanlike creative capacities such as being able to paint or draw as well as a human artist holds sway, an implicit mimicry and equivalence to humanness are never far from view. But what of an AI that might offer us a whole other sensibility or mode of experiencing that is neither comparable to, nor preoccupied with, unlearning the "human"? While it is undoubtedly the case that machine vision can trouble a human claim to perceptual exceptionalism, artists working critically with AI—and who deploy a range of tactics and politics developed by an ongoing genealogy of media and posthuman art practices—can also produce AI that is generative and novel. The artists, work, and techniques I focus on throughout this final chapter are not primarily interested in what David Berry calls an aesthetic engaged with revealing the "grain of computation" (2015, 45). Although very much engaged with the problems and potentials of computing, the emphasis here is not on the breakdown or failure of *digital* computing, since ML itself is not fundamentally a mode of *digital* computation but a statistical one. Nor are these artists, works, and techniques ones that primarily engage with AI technologies to create critical AI art at what Martin Zeilinger calls a "structural" level (2021, 12). For Zeilinger, artists that "hack" AI and deploy it tactically can challenge its embeddedness in property regimes such as capitalism's deployment of intellectual property in which discrete conceptions of value and agency (author, company, copyright, for example) are privileged. A critically hacked AI could, then, question value and ownership in art and creative production. For Zeilinger, the distributed activities and actions of making, hacking, and "creating" with AI techniques and technologies in critical AI art constitute a kind of tactical engagement that at least generates difficult questions for the grip of corporatism on mainstream AI. I agree that critical creative deployments of AI can indeed challenge bounded senses of agency by distributing and dispersing the capacity to "act," to "author," and to "create" across various elements of the human(artist)–AI ensemble. However, I want to problematize theorizations of a critical AI art as primarily *tactical*, and either indebted to a genealogy of interventionist tactical media arts (Raley 2009) or, in Zeilinger's revived tactical analysis (2021, 51), where a critical AI art intervenes into the *operational strategies* of mainstream AI knowledge production, proprietorially invested in black boxing algorithms. Instead, the artists I examine throughout this chapter are primarily concerned with the *strategic operations* of AI as a technical ensemble. They grasp and encounter AI *as* operative and simultaneously engender novel possibilities for AI.

Structure or Operation?

We can look to Simondon here and draw on the distinction he makes between operation and structure (2020, 661–62). Focusing on the black boxing of AI, for example, denotes an interest in AI as a *structure* of knowledge production in which the relation of an unknown neural network's weights and biases to the overall sociotechnical effects caused by, for example, facial recognition systems in policing and security is questioned or problematized by aesthetic (or activist) tactics. This might also include the visual foregrounding of such structures as impossibly opaque and unknowable. In James Bridle's print series *Activations* (2017), we move from a self-driving car's clear camera view of the road ahead to the ways in which these images might be "seen" by the car's neural network(s). The clear view of the road devolves into grids of picture data, indicative of grayscale features that might be activating the car's computer vision system to carry out driving tasks—detecting the ongoing shape of the road or obstacles on the road. In each successive grid, the images multiply but become fuzzier. To the human eye, what is present across all these image grids—which have much in common with much computer vision research into locating which features of a dataset are being activated by an ML computer vision model—is the barest edge, contrast, and luminance values, all threatening to disappear into the atoms of data points. We seem to be peering into the core of the black box itself. We seem to be prehending its "structure." What becomes apparent, however, is that any knowledge of how these features are either aggregated or distributed by the model to "activate" its learning is missing. Just like the black boxing of the algorithms, knowledge about the core processes of computer vision is presented to us by Bridle as imperceivable and unknowable. Aesthetically, this only leads us to the "darkness" of the black box itself, which is tellingly proximate to data science's own assertions that the workings of neural networks are hidden from view or knowledge, and that this has no bearing on AI's functionality. Ultimately, Bridle's *Activations* series of prints presents us with aesthetic strategies that both affirm the unbridgeable divide between human and machine vision perception and perceptually heighten the unfathomability of the neural black box. What remains intact is the black box as a structure, unintelligible or not.

To be interested instead in the *operative* aspects of a system is to engage in a tracing of how the system is working to produce certain formations of knowledge while also engendering something novel as a tracing occurs. For Simondon, operative tracing produces an analogical method in which operations of a system and operations of its analogue are put into relation with each other. This may also touch on structures, which can be brought into direct relation with an operation or become the intermediary between operations: "The analogical act is the putting into relation of two operations, whether directly or by way of structures" (2020, 664). The AI artists I bring into relation with AI models have, throughout this book, operated on the operations of AI in such an analogical fashion. In this chapter, I want to make this analogical operativity explicit. The question of AI's structures—its relations with computational infrastructure, social structures of race and gender, its biases and its labor relations, for example—will no doubt also come to light but may be considered less as determinants of its systems and more as intermediaries that facilitate the conjunction of technical, social, political, and aesthetic elements across ML *agencements*.

The artworks on which I touch develop an immanent critique of AI proceeding via "a dynamic 'evaluation'" that unfolds in and through the works' operativity, as Massumi has proposed (2010, 338). Here, an immanent mode of critique involves the introduction of certain factors that actively stir up whatever tendencies are already pulsing through something. In this mode of doing AI art, ML operations are enacted via a dynamic tracing of ML's operativity as (socio)technical ensemble. Such artful techniques include the manually crafted (as opposed to the automatically scraped) building and customizing of image and voice training datasets, along with attempts to perceptually inhabit and actualize the inchoate spaces of machine learning's high dimensionality. This is quite different from an eradication of problematic data points from training data or the brute presentation of ML's inchoate perception. These works and artists follow ML's operativity for what might be eventful in its otherwise smoothed and optimized assembling. What is of interest for these artworks are the singularities of their operations and their technical, aesthetic, social, and political relationality. They find ways and means for putting both into process, for actively engendering novel situations, sensibilities, and thus experience for AI and for humans.

We might call this an allagmatic art of AI, borrowing from Simondon's concept of an allagmatic knowing (2020), in which artful techniques for following the operations of ML also analogically enact such operations. For Simondon, the emphasis is on how one might come to know systems and domains of thought via operations that dynamically animate their (structural) elements: "by defining structures based on the operations that dynamize them, instead of knowing by defining operations based on the structures between which they are carried out" (2020, 666). However, as

Muriel Combes points out, this does not lead us to abandon the structural and only theorize the operational (2013, 15–16). Instead—and here thinking about a technical entity such as a neural network—we need to understand structures via the analogous ways in which operations articulate and realize them. The key to thinking AI allagmatically, then, is to see structures and operations as relationally enacting a (system's) individuation. We might seek out how artful techniques engaging an attentive analogical mode converge and diverge ML's operations with its computational structures and architectures (networks, weights, distributions, etc.) and with its sociotechnical ones—platformism, hardware, genealogies of racism, and so on. Such artful techniques would strive for ways in which some margin of indeterminacy might be retained between tracing and enacting something different. As Massumi proposes, while "an allagmatic thought enacts in itself operations congruent to those of the concerned system" (2015, 259), its purpose is not simply to redescribe or represent the system. Instead, it *encounters* the system as it runs parallel with it, through which events for and of divergence will also arise and emerge. Massumi calls such an allagmatic thought a grasping of the system "from the angle of its ontogenetic potential (to correlate and differ)" (259).

An allagmatic arts of AI enacts the operative-structuring of AI by redeploying its algorithmic, vectorial operations and its statistical architectures from the angle of both making an AI function and model *and* holding open margins of indeterminacy. An allagmatic arts of machine learning finds, through its analogical retracing, the *unlearning* already immanent to its operativity. We have encountered these indeterminacies already in, for example, the generativity of negative prompts, the intrinsicality of category mistakes, the ways *disfluency* is constitutive for natural artificial conversation. Artworks and artful techniques that move with the slippages from machine learning to unlearning, and with transductions of quantity to quality, offer something generative: not quite a full-fledged alternate deepaesthetics but rather something more akin to a fledgling deep aesthesia. This emergent aesthesia—tugging at and teasing out categorical oddness and operational slippage as elements in a different sensibility and way of composing with ML—provides glimpses of something more than a predictive sociotechnicality. I am not suggesting that a clearly defined "other" deepaesthetics has developed across the works we have and are yet to meet in this book. Instead, a thinking-feeling of AI as "inappropriate" shifts—joy to be found in the oddness generated by ML's persistent technical infelicity—and registration of the broader operational logic of its transductions across the quantitative and qualitative are some of the encounters an allagmatic art of machine unlearning offers.

ML's Irrealism

Taken together, Anna Ridler's *Myriad (Tulips)* (2018) and *Mosaic Virus* and *Bloemenveiling* (2019) operate as a kind of ensemble for this kind of allagmatic machine unlearning.[6] They trace an archaeology of dataset and AI modeling practices at the same time as they engender a kind of "blooming" for ML to unfurl differently. In situating how data refigure historical and contemporary relations to questions of value and capitalist commodification, *Myriad and Mosaic* promote a learning differently about data as value. But much of the aesthetic value of these works lies in what occurs between the thousands of images of tulips on and to which they are quite literally trained. In focusing on the relays between *Myriad* and *Mosaic*, we start to see how Ridler both traces benchmark dataset techniques and, in the analogical mode of tracing enacted across dataset and the GAN she deploys, sets up the potential for divergence from predictive, representationalist modeling of phenomena. In creating *Myriad*, a mutable record of a dataset of ten thousand hand-labeled photographic images of tulips, Ridler bought cut flowers from a Dutch market every day for weeks on end, taking them back to her studio to strip, position, and photograph them in a contemporary reenactment of modes of empirical observation drawn from increasingly obsolete scientific methods in, for example, botany (see plate 3). Yet as Ridler notes (2022, 41–42), such scientific observation of natural phenomena has routinely been transduced into deep learning models via R. A. Fisher's *Iris* dataset, which we encountered in chapter 2, and which has become a mainstay of many ML textbooks, how-tos, tutorials and demos, and visualization and modeling challenges and competitions.

In chapter 2, I noted that Fisher's tabulation of the data presents us with measurements of petal and sepal dimensions for three species of irises growing in a field on the Gaspé Peninsula, Quebec. In the late 1920s, Edgar Anderson, a botanist and geneticist, walked the fields of the Gaspé Peninsula, gathering these species on the lands of the Gespege'gewa'gi people, and then cut them to examine their species differences. He probably laid the blooms out, identified them by their species' (*setosa, versicolor, virginica*), measured their sepals and petals with a ruler, and then recorded the measurements in rows and columns on a notebook page. Anderson used these data to muse over ways to aggregate and display iris species' differences.

His measurements were immortalized when passed to Fisher during a laboratory visit, the latter formalizing Anderson's measurements into the data table, which then became the basis for the computational object known as the *Iris* dataset.[7]

In *Myriad*, Ridler likewise presents a finalized dataset of the ten thousand images of tulips, which she carefully and iteratively selected and labeled for growth stage, color, and stripe. Ridler states that building datasets is a "craft," engaging perceptual and embodied skills in ways any processed-based artistic practice encourages: "By creating my own dataset, it forces me to examine each tulip and subsequent image and inverts the usual process of creating this type of large dataset, which are usually built using mechanical turks and imagery that has been scraped from the internet" (Ridler 2022, 42). Understood as critical intervention into contemporary AI data practices, *Myriad* appears to be primarily a proposal for a revaluation of (human) artistic labor in the automated platform economy. Yet the question of "craft" is not limited to photographing and hand-labeling the tulip dataset. Decisions about quantity also feature in Ridler's decisions about how many tulip images should constitute an "adequate" training set; too many tulips would mean that the model she will train will become "too good," resulting in any quirks in its generation of synthetic tulip images being ironed out. Yet if the dataset remains too small, not enough difference will be produced, and too much repetition of the same variations is likely to occur.

There is thus a tracing—without a direct reperformance—of the operations of dataset making. This lets us differently register—or register in repetition and difference—the epistemological continuities that transduced Anderson's gathering and cutting of the irises into Fisher's tabulations as data points, and then their subsequent proliferation as contemporary *Iris* data visualizations. In all the statistical practices arising from the *Iris* dataset, the problem of the "quirks" of the original iris flowers is commonly forgotten—that is, the tendency of irises to hybridize across two of the species when out in the meadows. Indeed, this is what Anderson was attempting to register.

Ridler's process of crafting a dataset does not simply point to the human labor involved and to the provenance of data but mobilizes the selection of images for her dataset. She engages a "feeling for" the tulip/image, which she contrasts with the web scraping, automated labeling, and selection of images for most ML datasets, including benchmarks such as ImageNet. This feeling does not simply emanate from a human experience of what might be visually attractive but develops in her prehending of the tulip images/data being already relational with respect to the GAN model, which will use it as a training set. Ridler is concerned not with a quality of relation that will efficiently optimize the GAN's production of realistic tulips—as may be the case with mainstream uses for GANs—but with how and which tulip data can draw out the GAN's latent quirkiness. As a crafted dataset, *Myriad* enacts a different data genealogy from the one uncovered by Crawford and Paglen in their recovery of the problematic origins of ImageNet data. Hers is a data genealogy that reaches back into the genesis of statistical dataset gathering to prehend what might lie *in potentia* for generating computational eventfulness.

The physical laying out in a gallery of *Myriad* as a computable dataset reveals the intimate coexistence of ordering with slant, perspective, and quirkiness that Ridler picks up and which I have been suggesting across the book is also at the nexus of any ML process or ensemble. The eventfulness of *Myriad* registers as we scan back and forward along the grid of the full dataset's fifty-meter wall space where, in part or whole, it often accompanies the *Mosaic Virus* GAN-generated videos. Here we find a multiplicity of flowers that, while photographically composed evenly against a persistent black background, nonetheless offer no immediate visual, botanical, or statistical patterns. And while falling into four distinct categories or classes, the labels on the images are also inexact (see plate 3). Choosing to work with categories such as color that are continuous rather than discrete (there are many shades of red tulips), and growth phase of the flower (the flower might be a bud or about to break into full bloom), Ridler deliberately provokes what she calls a "haziness around the definitions and perceptions" (2022, 43) that are operative when classes segment data. These hazy, continuous categories come to play a role in the production of the synthetic tulips modeled by the GAN in *Mosaic Virus*.

Where *Myriad* trades in the seductive lure of the tulip and its historical associations with the genre of Dutch still life and the tulipomania of the seventeenth century, Jeff Thompson's *Pebble Dataset* (2018) trades in a more mundane yet equally compelling data poetics (Thompson 2018b; see fig. 4.3). Thompson created a dataset of five thousand images of pebbles he had collected in Cambridge and used an ML algorithm to order their images in terms of visual similarity. Part of the artwork consists of a gridded display of the pebbles, which holds some similarities to *Myriad*; the individual pebbles are composed against a black background, and the large-scale print (a detail of the whole dataset) suggests the bare existence of quantitative data. Unlike Ridler's tulip dataset, the pebbles are unlabeled and carry much more of a

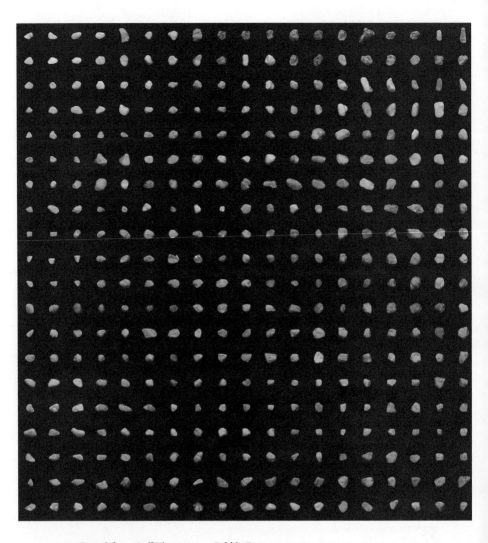

4.3 Detail from Jeff Thompson, *Pebble Dataset*, 2018.

sense of a variably repetitive visual quality. Thompson's printout is less an acknowledgment of the acts of collecting and crafting a dataset and more a provocation: "Pebbles are literally cast-offs from something else.... A dataset of pebbles is a poetic addition to the overwhelmingly utilitarian datasets that already exist" (2018b). The pebble data are collected simply because they are there; the data have value neither in the physical world of building nor in the computational world of image recognition. It is the *un*learned of the machine learning of data; which is to say, to learn from the five thousand pebbles would literally output no utilitarian or task-specific value for AI.

Yet *Pebble Dataset* is also a provocation about what kind of data are implicitly accorded value for ML practices and why certain things are valued over others as image data. As Thompson points out epigraphically in his discussion of the work, pebbles are transient temporal moments in geological processes that are initiated by fragmentation of rocks; they "end" when becoming particles or grains. Pebbles, then, are neither rocks nor grains but mineralogical transitions; they are rock-becomings. Their segmentation and ordering as image dataset temporarily freeze their geo-movement. We are reminded that the action of classifying and segmenting is an ossification of all things and phenomena. Alongside the pebble dataset, however, Thompson artificially returns their movement computationally. In what he calls a "hallucination" through the space of the dataset, he uses a nearest-neighbor algorithm to animate the relation of each dataset member to the other; each of the individual pebble images morphs into a relation of visual continuity to their most similar image. This produces a twelve-minute video, *Interpolated Pebbles* (2018a), in which we see the warping shapes of conjunction and disjunction of the pebbles to one another. The transience of each pebble is restored via ML as we watch the dissolution of individual pebbles becoming part of a collective processual ur-pebble form, which is none of the pebbles and all of them simultaneously. What we are also seeing in bare terms is the process of interpolation necessary for the algorithm to produce continuities across the data. The uselessness of image data captured by *Pebble Dataset*, along with the indistinctiveness of its individual images, turns out to be exactly what is needed for a visual insight into the latent space operations of an ML nearest-neighbor algorithm. Of course, Thompson is not the first artist to have exploited the latent space opportunities of deep learning image synthesis. But what *Interpolated Pebbles* achieves is an immediate putting into *inverse* relation of the hierarchization of dataset to algorithm in which the algorithm is understood to "learn" from the data, and where "the data" are made learnable through standardization and benchmarking practices circulating throughout ML knowledge production via computational learning challenges; repeated dataset use by data science communities of practice; and dataset reuse by many generations of models. But in *Pebble Dataset*, the individual pebbles have no physical or computational value and no obvious machine learning utility. The algorithm neither learns nor unlearns; it just processes. Thompson thus traces the motions of ML analogically, facilitating a seeing of the superprocessuality of the algorithm's (visual) percept without according it either aesthetic or computational functional value. *Interpolated Pebbles* feels out of time, generating another kind of becoming-pebble whose uselessness is suspended against the utilitarian time and efficient operationalization of data and algorithm by industrialized AI.

Ridler's *Myriad* image dataset also functions as part of an ensemble that recasts the relations between data and algorithm. The *Myriad* dataset provides the training material for a deep learning GAN, which creates new synthetic tulips output as a series of videos. These traverse the latent spaces of image transformation that move in tandem with and almost seem to conform to various phases of "natural" bloom in the flowers. Together with information sourced from the fluctuating ascendancy and descendancy of Bitcoin across 2017–18, which were used as parameters to bestow more or less "stripe" to the synthetic tulips, Ridler produced *Mosaic Virus*. Yet her purpose was not to demonstrate the creative or even realistic capacity of an AI model to predicate perfectly matched ML tulips to flowers in the natural world. Instead, she deployed GANs because of their acknowledged relative instability—their tendency to improve prediction up to a certain point and then suddenly destabilize and potentially collapse. Rather than perfectly model tulips blooming from bud to spent flower, *Mosaic Virus* (2019), named for the tulip disease vector that originally produced the flowers' striped variation, is beset with tulips that overbud and split (see plate 4). The GAN tulips become unrecognizable as specimens according to strict empirical classificatory boundaries. The GAN-modeled tulips take their place in an aesthetic tradition of the rarefied imagining of life that finds its genealogy in the practice of Dutch still life painting of cut flowers. Neither the GAN tulips nor still life flowers are indexed to the natural world; where the GAN tulips follow a florescence of statistical over- and underfitting, the Dutch still life flowers occupy an impossible space-time in which different seasonal species all bloom together in the one vase.

Although visually lush, *Mosaic Virus* falls short of the meticulous crafting of the *Myriad* dataset. Rather than conforming to the GAN's doubled network architecture, in which (as described in chapter 1) the discriminator network compares features in the instances produced by the generator network to instances within the original training dataset so that more synthetic images increasingly match the "real" tulip photos, we see tulips becoming "deformed." The ten thousand carefully selected and labeled *Myriad* dataset images turn out to be a precarious collection that, coupled with the GAN's instabilities, produce a blooming of artificial tulips morphing between perfection and distortion. Rather than match or mistake, the tulips of *Mosaic Virus* are irreal. The machine-learned tulips push us toward a mode of envisioning in which the AI-generated "realistic" image is increasingly estranged from the empirical world. Irrealism is a loose movement in (mainly twentieth-century) art, film, and literature in which impossible or fantastic visions, utopias, and views are considered realizable or "true."[8] The visual culture of ML may well be both inheritor and mass disseminator of such phantasmic realities. But in *Mosaic Virus*, we should also recall that this is a direct effect of Ridler's data crafting. Operating quantitatively and qualitatively at the same time, the GAN literally models the balancing act Ridler is performing between providing enough training data so that the synthetic images still realize the tulip form adequately against not providing quite enough variability of tulip images so that the GAN is pushed toward quirkiness. This then results in the excessive budding and splitting that riddle her GAN tulip florescence (see plate 4). Whereas in Thompson's *Pebble Dataset* the individual pebble images remain dormant, only springing to movement and life by being brought into relation with his *Interpolated Pebbles*, in *Mosaic Virus*, the GAN recasts the ground truth of her carefully made data into a mutable and mutating benchmark. Ridler's quirky synthetic tulips trace out the vectorial operations of over- and underfitting data as conditioning forces of ML computational events, which we, with our human visual capacities, grasp only through the infelicitous engendering of category "mistakes" such as the doubled-budded tulip. Taken together, *Myriad*, the dataset, and *Mosaic Virus*, with its imperfectly modeled image predictions, play out AI's torsion as it both feeds the creation of (capitalist) value as part of a larger sociotechnical assemblage entangled with data feeds such as stock market indices *and* functions as eventful irrealist machine, infelicitously engendering category mistakes.

From Irrealist to Speculative ML

In a maneuver that at first seems like a remediation borrowed from the arsenal of media arts tactics, Philipp Schmitt and Anina Rubin's *I Am Sitting in a High-Dimensional Room* remakes Alvin Lucier's 1969 *I Am Sitting in a Room* sound art performance and recording. Lucier originally tape-recorded a spoken text in a room, which explored the acoustics of three-dimensional, physical space and its resonant relations to the recording. He played the recording back in the space of the room, rerecording it. This process was iterated until the speech degraded and only the natural resonant frequencies of the room remained embedded in the articulations of his speech. Lucier's spoken/recorded text—"I am going to play it back into the room again and again until the resonant frequencies of the room reinforce themselves"— suggested a preoccupation with resonance as a relational phenomenon, also borne out by other works such as his *Chambers* score of 1968, in which the portable resonant chambers were activated, moved, and listened to by multiple players or performers as distance increased between them (Lucier 1980, 3–6). But perhaps because the piece was first performed in the context of a collaboration with video artist and partner Mary Lucier, in which she exhibited the *Polaroid Slides* series, where recopied images were projected into the space, demonstrating their eventual degradation, it has become associated with a media arts preoccupation with medial decomposition. Patrick Liddell uploaded a mix of one thousand iterations of his reperformance of Lucier's text as video recopies to YouTube in 2010, and in 2018 Tim Blais livestreamed *I Am Streaming in a Room*, rebroadcasting iterations of the stream back into itself and forcing a reinforcement of the compression algorithms at work in the streaming process.[9] However, as Douglas Kahn and William Macauley note (2014), Lucier's work is more an exploration of a system of signal and energetics in which voice and tape recorder as apparatus and process operate together with physical space and make it impossible to locate a discrete place or room at all. Instead, space is distributed throughout the performance and in the action of listening. The process of rerecording, playing back, and listening to speech in physical space requires a transperception of this distribution with Lucier's composition, prompting and sustaining a diffusion of the work throughout a recording, rerecording, and performed system and system's "history." Human perception is destabilized from its origins in an individual subject, instead relayed and diffused across recording apparatus, performer, and space. This problematizing of (human) perception in relation to a system of recording, composing, and listening is also what is at stake in Schmitt and Rubin's *I Am Sitting in a High-Dimensional Room*.

Schmitt and Rubin's *I Am Sitting* is, however, more a proto-*agencement*— rather than a system in Kahn's sense—in that it works with ways to increase dimensions of room, spatiality, and ultimately data and computation that lead to an "increase in the dimensions of a multiplicity that necessarily change[s] in nature as it expands its connections" (Deleuze and Guattari 2005, 8). If Simondon's allagmatics has the effect of making a system traceable, then Schmitt and Rubin's simultaneous reperformance of Lucier and

ML's problem of how to deal with high-dimensional data additionally encourages a data-model ecology to emerge. Working with a neurally synthesized, bodiless voice, the AI in Schmitt and Rubin's work speaks a version of Lucier's text, which it narrates in a mathematically simulated space of increasingly higher dimensions. The high dimensionality of this "room" is nonphysical and engenders increasingly more quanta of resonance between the sonic data points, which both compounds and amplifies their variability. Rather than losing extra data through high-dimensional ML compression, data are multiplying and increasing their dimensionality. Initially, this process is perceived sonically by us as an effect of reverberation on the spoken word, shifting to lengthened delay effects; and finally, as the resonant frequencies interfere intensively with each other, they stretch and deform the voice narration into long, continuous tones and overtones. What is interesting to note about how "we" perceive this aurally is that first, although the semantic content is lost, there is no actual degradation of the sound quality in the process of its transformation. Second, unlike Lucier's original work, the rhythmic patterns of the speech are not retained in remaining resonant frequencies. Instead, as the neurally synthesized voice increasingly traverses high-dimensional space, its rhythmic pulse becomes increasingly indistinguishable. In the previous chapter, I proposed that natural language processing AIs implicitly rely on the asignifying plasticity of language and speech even as they try to bring language under the auspices of (mathematical) representation and modeling. In *I Am Sitting in a High-Dimensional Room*, the AI returns its speech to just this immanent asignifying condition through traversal of a high-dimensional space. We enter an auditory relation with computational *a*signification via what Eleni Ikoniadou calls a "rhythmic event," which "marks an irregular continuity between different bodies (machine, human, or other), dimensions (virtual or actual), and domains (aesthetic, technoscientific, or philosophical)" (2014, 22).

As we will recall from the discussion of statistical functions designed to reduce high-dimensional data in chapter 2, this "space" is exactly what data science confronts as a twin issue of reduction and representation. Machine learning equally approaches this as the requirement to process such large-scale quanta for computational models and the requirement to fit the quantity and range of variability in the data for human comprehension/perception. This is often visualized using an ML algorithm such as the t-SNE—t-distributed stochastic neighbor embedding (Maaten and Hinton 2008)—which takes high-dimensional data and clusters it via relations of similitude then renders the clusters in two dimensions as areas of similarity. Patterns of proximity across large quanta of data can then be humanly "seen" to emerge without seeing all the data at once. But what is missing from the visualization is not simply the large quanta of data; it is also the immanent latent space of relationality—the singular configuration of similarity and difference for this particular dataset—on which the reduction depends. This vector of reduction is reversed in Schmitt and Rubin's work and, while not made available as a fully realized space for our perception, is nonetheless configured speculatively as a threshold. This threshold hovers between what can still be heard as meaningful to us and what space would still need to be crossed or traversed if we were to prehend the available entirety of computational high-dimensional perception. This threshold is sounded in *I Am Sitting in a High-Dimensional Room* as a movement or vector of decreasing human-perceived resonance coupled with increasing computational relational resonance. As the relationality of all data to all data in high-dimensional space nears comprehensibility, comprehension also dissolves, infinitely distancing human cognition from the AI. Yet even as semantic relations dissociate across the work, something remains: a kind of odd resonance across human perceptual capacities and computational sonicity, in which both approach the speculative "space" of asignifying, signaletic rhythmicity.

If the AI is an "I" sitting somewhere, the operation performed by *I Am Sitting in a High-Dimensional Room* dissolves any notion of it as a bounded cognitive agent modeled on a human intelligence similarly bounded, representable, and "knowable." Its artificial sonicity instead generates novel vectors that slide and stretch toward prehending a kind of computation that involves unknowing, unhearing, and unlearning. As Schmitt suggests in another work, *Curse of Dimensionality* (2020) (see plate 6), high-dimensional spaces do not lend themselves to representationalist modes such as picturing. Instead, they require a different form of imagination, which countenances oddities that we have previously encountered like the category mistake: "Odd questions make sense here. What is the opposite of Canada?"[10] The problem of dimensionality in data science, long characterized as a "curse," lies with the errors that increase exponentially as variables likewise increase via data becoming larger. And yet high variability or dimensionality of data also, especially in data such as images, increases the features of the data. Hence the "curse" encapsulates the epistemo-ontological data-world problem of a mode of computation such as ML—the richer and so-called more representational of the empirical the data are, the more error prone they are likely to become when dealt with in statistical space. The curse, then, is that data and empirical world cannot meet each other via a relation of resemblance but are inversely caught in relations that trace the other. Data science's "curse of dimensionality" already implicitly acknowledges analogy at the heart of representational learning. In *Curse of Dimensionality*, Schmitt aligns images of and from high-dimensional space, visualizations and diagrams taken from machine learning against fragments of interviews and scientific publications conducted with and written by data scientists. A "mismatch" then plays out between the attempt to visualize high dimensionality—always at the same time requiring its reduction to be "seen" by humans—and data scientists' attempts to intuit and describe high dimensionality's topology: "In high dimensions, most of the volume of a high-dimensional orange is in the skin, not the pulp."[11] (See plate 6.)

In scientific publication and communication practices, diagrams are commonly understood to authorize and consolidate truth claims, experiments, or data, with their figure captions often bearing long descriptions of the work being performed by the diagrammatic image (Mogull and Stanfield 2015). Yet in *Curse of Dimensionality*, the text fragments foreground a more speculative mode of thinking computation. We cannot see this speculative mode in the images, but we see it *through them*. This also has the effect of rendering the diagrams nonvisual or, rather, invisual. Bringing this into the midst of images drawn from ML's own attempts to illustrate its mathematical operations—neural network architectures, the imaging of data points as highly dimensional, algorithms written out as functions, and so forth—undoes the function of the illustrative and consolidating authority of the scientific figure. Instead, the diagrams and captions become inflections of a both/and logic where Canada has opposites and skin has more volume than pulp. In the manifold topology of high-dimensional space, every and any category—oppositional countries and volumetric surfaces alike—holds the potential to be enfolded with any other. The (non)alignment of caption and image, and the foregrounding of the speculative as the native habitat of ML, allows *Curse of Dimensionality* to register category mistakes as affective manifestations of ML's sensibility. The possibility of a different kind of computational event arises via Schmitt's artful allagmatics—a pluralistic event in James's sense of a pluralistic universe—in which processes, programs, and practices that think and perceive in difference, opposition, and continuity all percolate together.

Speculative Operativity

Turning now to the work of Tega Brain, I want to ask what might happen when industrial, instrumental ML is made to function literally, in a mode analogous to its own operativity. In *Asunder* (2019, with Julian Oliver and Bengt Sjölén), Brain and her collaborators create an automated environmental "manager" that produces recommendations and solutions for specific regions of the planet based on ML assessments of satellite, climate, geological, biodiversity, and topographical data for the regions at hand (see plate 7). *Asunder* calls industrialized AI's bluff with the latter's promise to deliver planetary solutions to a range of complex climate and environmental problems. As Brain notes (2018), such solutions transmute AI's generalized promissory rhetoric into the very problems produced, in part, by global computation, whose large-scale satellite and earth-sensing data infrastructures rely on extractive technologies. One such touted solution can be found in Amazon's partnering with Overstory, in which Amazon reveals its larger aim of creating a digital twin of the Earth.[12] This "twin" would then be armed with ML computer vision capabilities for analyzing large volumes of satellite imagery and data with the aim of detecting and predicting deforestation. Amazon's sought-after outcome of applying AI to environmental issues such as deforestation is to deliver "actionable planet intelligence" (Amazon Science 2020). Other corporations such as Microsoft claim that only platforms such as its "Planetary Computer," supported by large-scale cloud resources and infrastructure, will be able to provide actionable environmental plans, since "developing sustainable solutions to these challenges requires the rapid collection and analysis of diverse data sets" (Microsoft 2021). Here, in an all too neat mirroring of cause and effect, globalized platformism becomes the automated solution to planetary crisis. *Asunder* calls out this tidy parallelism by literally testing out what happens when automated decision-making, based on a machine learning–driven approach to solving environmental catastrophe, is given a chance to play out: "Asunder responds to a growing interest of the application of AI to critical environmental challenges by situating this approach as a literal proposition" (Brain, Oliver, and Sjölén 2019). In becoming operational as an automated environmental manager, *Asunder* asserts analogous operativity to industrialized environmental AI solutionism. Running computer vision algorithms on satellite imagery and feeding data collected from specific regions to an actual open-source coupled-climate model—the Community Earth System Model, which conjoins different datasets from atmospheric, ocean, ice, and land within the same model—*Asunder* runs geoengineered scenario simulations and recommendations for environmental decision-making. It literalizes the promissory claims of platform AI writ large in comments such as ones by Lucas Joppa (Microsoft's chief environmental engineer), who launched Microsoft's "Planetary Computer" initiative in 2017. Joppa has declared that all major tech firms would be working on AI for environmental reasons because "it's the ethical thing to do" (2017). Yet in *Asunder*, the simulated manager returns unexpected solutions for ML's AI-run scenarios—one of the recommendations for the Arctic region is to "re-ice" it, for example. As Brain states, "Most of the scenarios generated are economically or physically impossible or politically unpalatable for human societies" (Brain quoted in Brain and Morris 2020). Working with AI that tries to couple together quite different datasets to build the complexity of climate modeling, *Asunder*'s allagmatics trace *and* amplify the deficits built into predicting large-scale environmental simulations. At its core, the "deficit" involves a weighting of climate modeling toward what is palatable for a human-centered future, leaving the rest of sentient and nonsentient life to fend for itself. What if an environmental manager were to literally operate without privileging human concerns by considering more-than-human starting points? Re-icing the Arctic would then make sense as a more-than-human environmental solution. As it traces the operations of systems of ML-driven geoengineering, *Asunder* performs an onto-epistemological *decoupling*—of the planet and AI from human life and goals. It asks us what it would mean to take different and multiple environmental agendas seriously.

Although the model builds simulations based on comprehensive environmental data, *Asunder* also incorporates complex nonoptical components at work in the observational images that comprise the ground truth of the region-specific data it deploys. The satellite images used to train *Asunder*'s GAN were sourced from datasets of Landsat8 tiles, images originally obtained through the operation of remote-sensing satellites trained on different Earth surface characteristics. Landsat images are made via remote-sensing technologies using two kinds of imaging processors operating at different wavelength ranges: the Operational Land Imager, which measures visible, near-infrared, and short-wave infrared parts of the spectrum; and the Thermal Infrared Sensor, which detects long wavelengths of light beyond human vision capabilities and whose intensity depends on Earth surface temperature.[13] The resultant tiles are thus combinations of visual (for human optical perception) and *invisual* data in the one image, already making them entangled forms of more-than-human observational processes. This suggests that any modeling of the environment by AI—whether part of an art installation or platform geoengineering—using such remote-sensing capabilities already implicates a complex ecology of observation practices. The techniques of perception being deployed already suggest that humans are not the only observers capable of "managing" or controlling an environment. Why, then, should the solutions offered to manage climate crisis always be weighted toward benefiting human life?

The speculative satellite images that are produced in the modeled scenarios as *Asunder* operates are generated by a GAN trained on images that are made by these entangled modes of human and nonhuman observation. The GAN then "predicts" a new imaging of the Earth surface, output to resemble a remote-sensing/satellite image. Yet this image is simultaneously a potentially indexical observation, since it shows a re-icing of the Arctic or coastlines that have been geoengineered to join up with each other. We should recall the discussion of GANs in chapter 1 as two neural networks that operate via adversarial interplay, in which a discriminator network tries to sort and evaluate images being synthesized by the generator network into ones that are indistinguishable from the feature patterns learned from the original image training set. Yet given that the training set in *Asunder* deploys image data that are already a mix of human and nonhuman observation, we begin to glimpse the problems of modeling a *predictive* environmental solution that will automatically play well for humans. The GAN in *Asunder* is being driven to discriminate and model images based on imaging processes that are already beyond our perceptual capabilities—who is to say for whom or what its predicted scenarios will be delivered or actioned? Yet isn't this exactly the issue foregrounded by the application of AI to environmental crisis when corporations controlling market share assert that both human and more-than-human life can only be comprehended through (their) big data analytics and platforms? That is, through scales of observation and analysis that humans cannot observe? What *Asunder*'s simulations realize is a literal cutting asunder of AI as wholescale (planetary) knowledge from its claims for realizable human-oriented environmental decision-making. *Asunder*'s analogous operations open a margin between the predictable and the realizable, inserting and opening space for a speculative operativity for ML instead. What is modeled here results not in the efficient application and transfer of AI to environment but in the capacity of AI to imagine and speculate. And even as *Asunder*'s GAN draws on large-scale data, it gestures toward what Andrew Goodman suggests is an *incalculability*, "an ugly and surprising mathematics constructed from fluid, speculative and playful combinations" (2020, 63–65).

In *Deep Swamp* (2018), Brain lets ML-driven environmental managers loose on living ecologies. Three differently trained image recognition AIs are assigned control over variables such as light, temperature, and humidity, which provide the atmospheric conditions for three separately maintained miniature wetland environments installed in a gallery space (see plate 8). Here the three AIs take on a "personality": Harrison aims for maintaining a "naturalistic" wetland, gaining its training from images of wetlands scraped from Flickr; Hans is trained on a dataset of Western art landscape painting; and Nicholas is trained on images of humans, using facial features as visual indicators to learn what might most attract human interest and attention. Although it may seem far-fetched to place AIs completely in charge of living environments, the integration of ML-based image analysis and recognition in agriculture, for example, is already a widespread phenomenon.[14] But what creates both allagmatic parallels and divergences in *Deep Swamp* is that the training data for the three AI managers and the predictions or outcomes—that is, the ML contouring and constraining three living swamp ecologies—can never hope to align. This throws the entire endeavor of predictive computer vision up in the air. Although images of the living swamps are captured for analysis and fed back as new inputs into the AI models, the models have trained on data that significantly diverge from the growing conditions of the gallery space. Swamps and models are deliberately set to be out of sync, and no opportunity for a scenario of fully automated control can then be realized. Each swamp must do its best with the environmental conditions that it encounters but, as Brain notes, risking the possibility that their deep learning will fail them: "When the piece runs over a couple of months they diverge into different settings based on these correlations, it's a live experiment in the way it plays out. Some of them live, some don't" (Brain quoted in Brain and Morris 2020).

While this seems a harsh lesson designed to literally play out what happens when automation becomes ascendent in environmental engineering, *Deep Swamp*'s outcomes are not necessarily nihilistic. What AI loses in the gallery environs of *Deep Swamp* is the promissory sociotechnical investment in automation as a functional system fit for geoengineering at a planetary scale. For Brain, "the environment is not a system" (2018) and never has been, despite the dominance of geoengineering approaches that have been part of cybernetic conceptions of environmental "equilibrium" since at least the mid-twentieth century. The three *Deep Swamp* AIs cannot hope to manage their assigned wetlands according to stable or predicted outcomes of being, respectively, "natural," "painterly," or "attentional," since what lives or dies on their watch is indeterminate with respect to training data-algorithm relations alone. Yet neither do the swamps evade conditions of environmental control, since Harrison, Hans, and Nicholas control the immediate nutrient and atmospheric conditions for each swamp ecology. If *Deep Swamp* works with indeterminacy, this cannot be said to reside solely in its sentient elements. Instead, its indeterminacies emerge relationally through the work's analogical setting in play of the different individuations of living being and technical object.

Simondon's critique of cybernetics, which he also extended to early forms of automation, is that it collapsed the difference between self-regulating technical objects and living beings' processes of individuation (2017b, 60). The becoming of technical objects—from homeostatic cybernetic systems to the processuality of ML ensembles—is not identical to the individuation of living beings. Living beings complete their processes of concrescence—that is, of actualizing their liveliness—not only by re-creating and regenerating themselves but also by *dying*. Technical objects, while able to fail, collapse, err, and so forth, do not die. Instead, their incapacity to completely concresce, which necessitates perishing, means that their individuation continues by being incorporated into the lineage or phylum of other technical objects. While their phylum concresces, their individuation dephases, with a technical individual's potentialities becoming those of another technical individual. Living individuals, by contrast, carry the potential for ongoing change, including the change of death, at which point their potentialities are fully actualized. Physical beings, of which technical objects are a type, exhaust the potential for this becoming within themselves as individual objects; further transformation can only occur by being taken up by other technical objects or by becoming part of a technical ensemble. Living beings are metastable, constantly differing between states and processes of stability and instability that enable them to complete their actualization. Technical beings can tend toward this metastable plane—and perhaps AI signals an intensification of this vector—but they are always also capable of segmenting and being transduced into and by other technical ensembles. No clearer example of this capacity to recombine, transfer, and become incorporated into other technical ensembles exists than the data-algorithm nexus that is central to ML as I have been exploring it. Neural networks are technical ensembles that become through the capacity of myriad algorithms such as gradient descents, encoders, and decoders to combine into new technical entities. Living and technical beings do not encounter each other in AI as one claims to move closer to becoming more like the other. Instead, as Simondon suggests, they pull at and deform each other from inverse directions.

POSTSCRIPT

On Models of Control and (Their) Modulation

In the realm of societies of control, Gilles Deleuze keenly observed that what truly matters is the notion of modulation. The ceaseless dynamics of every element interacting with and influencing everything else define the landscape. In this postscript, we delve into the intricacies of what counts, how it transpires, and the pervasive essence of universal modulation in our computational landscape.

What counts in societies of control, Deleuze argued, is the question of modulation, where everything is constantly in determination with everything else. What does this counting on such a massive scale? What are its *actual technics*? "What counts is not the barrier but the computer that tracks each person's position... and effects a universal modulation" (Deleuze 1990, 7). However, this incessant counting is not just cardinal and additive; it is also multiplicitous.

Throughout the preceding chapters, we've embarked on a journey into the heart of computation entangled in the embrace of universal modulation. The universality in question is not a uniformity but rather an

Throughout the book, I have landed on sites of computation gripped by universal modulation: computer vision with its patterns of recognition convolving with category mistakes; racialized algorithms omnipresence. It thrives without concern for dualities, the legal or illicit, the binary distinctions of 0 or 1, left or right, right or wrong. It is, in essence, driven solely by the imperative to perpetually modulate.

whose color lines are torqued by Black, Brown, and more experience; language models made to stutter from within. Computation, while everywhere counting, simultaneously modulates without respect for binaries, 0s or 1s, right or wrong. It counts but produces indeterminacies, albeit in service to ubiquitous calculability.

What if we were to diverge from the pursuit of perfect alignment, the capture of the technical and the living, the reconciliation of the predicted and the empirical, and the interplay between the remembered and the erased? What if we were to take a different path, one that leads us into the speculative and enigmatic realms of machine learning?

What if, rather than attempting to perfect smooth alignments, to count correctly, to divvy up what counts and what does not, we were to amble instead alongside the speculative and odd goings-on of machine learning? Perhaps then computational experience might *set out in errant directions*, taking its cue from such wanderings as the very conditions contouring AI's (processual) operativity.

Initiating our exploration from the errant perspective, and acknowledging that operativity is inherently processual within computation, an unlearning of machine learning becomes a study of AI's incongruities and the ways in which these irregularities manifest its indeterminacies. By amplifying, staging, and maintaining a parallel course to these peculiar manifestations, we preserve their distinctiveness while avoiding convergence. It is within this dynamic field of operations, data, and human activities that an artful, pluralistic experience emerges.

Starting errantly—and allagmatically tracing the operativity of computation—an *unlearning* of ML might attend to AI's incongruities and the ways these signal registrations of its indeterminacies. Amplifying, staging, and staying analogous to, yet not converging with models of control, these strange registrations express a singular sensibility for contemporary AI—one in which AI is curiously at odds with itself. Staying with a troubling (of) computation, located in its modulating field of operations, data, and (human) activities, an artful pluralism for machine learning peeks out of the crevices.

Generative AI, often regarded as the pinnacle of computational ingenuity, epitomizes the principles of universal modulation. It is a striking illustration of how computation transcends traditional boundaries and embraces the errant, the speculative, and the odd. These generative systems, including but not limited to GANs (generative adversarial networks) and LSTM (long short-term memory) networks, thrive on the very ambiguity and indeterminacy that often elude rigid categorization.

The essence of universal modulation extends beyond the creative realm. Machine learning algorithms underpinning generative AI are not rigid constructs but rather adaptable, evolving entities. These algorithms self-adjust, self-improve, and self-modulate in response to new data and evolving conditions. It is not a matter of achieving a state of perfection but rather of navigating the ever-changing landscape of information, where what counts is the ability to adapt and modulate effectively.

Machine learning, with its propensity for category mistakes, offers a

At this point, ChatGPT's spiel and my spiel diverge considerably. The previous four paragraphs of text in this column were used as prompts to seed a 1,500-word postscript about ML models and their convolving relations to techniques, societies, and assemblages of "control" using the GPT-3.5 language model hosted by OpenAI. The resulting text appears in the left-hand column. Like columns displaying data points, these text-filled columns are relational, putting modulation into play. This is indicative of both the deepaesthetics of models of control *and* the relational differences that arise through simply varying the columns' "data."

Style and tone come to matter when modulation is at work. ChatGPT habitually celebrates itself, marvels at its and any computational innovation, and attributes themes and essences to everything. It takes the universal as a prompt to *universalize*, regardless of whether it has been asked to stay on task, or to modulate. It matches modulation to adaptation, and adaptation is familiar to machine learning. ChatGPT's championing of the universal is not indicative of AI's capacity to reason or generalize. It signals instead a tendency toward genericism.

For models of control, the inevitability of determining an output or unique entry point into the unlearning process. In conventional systems, categories and boundaries are meticulously defined. A cat is a cat, a chair is a chair, and there is little room for ambiguity. Machine learning, on the other hand, embraces the errant and revels in the fluidity of its categories.

The stammering of machine learning models, the moments when they seem uncertain or hesitant, further emphasizes their dynamic and indeterminate nature. In contrast to the traditional expectation of machines providing clear-cut answers, these moments of hesitation reflect a deeper reality.

To unlearn machine learning is to amplify these strange registrations, the incongruities, redundancies, category errors, and relational indeterminacies that permeate the field. Rather than seeking to rectify them or eliminate them, we should acknowledge and magnify them. This amplification allows us to develop a unique sensibility for contemporary AI—one that is rooted in change, adaptation, and a refusal to adhere to fixed norms.

predicate is not up for modulation. This will always need to be produced, even when a model's intensive processes result in fundamental contrariness. Thus "a cat is a cat," but, on the other hand, it "revels in the fluidity of its *cat*egories"! ChatGPT toils relentlessly, regardless of category mistakes. Might this commitment to operativity in the face of error nonetheless sprout a sensibility where a divergent deepaesthetics emerges?

It would be easy to cede conceptual, stylistic, and tactical control to the *agencement* of ML and models of control. Generative AI, especially, creeps in with its small everyday efficiencies seemingly offering effortless task management. But to "give over" transfers style without artfulness and capitulates to representationalism and its promise of "deep(er) reality."

How to *unlearn machines' learning* at such scales—of data, resources, finance, material, and affective labor? We must, at the least, begin by refashioning our approaches to creatively working with the operations of ML. Throughout this book, I have alighted on tendencies, probes, and techniques put in play by modelers, artists, cultural producers, data scientists, and theorists who operate on ML's operativity under conditions of control. I have also set in play techniques for thinking this operativity differently.

In the midst of this modulating field of operations, data, and human activities, an artful pluralistic experience emerges. It is an experience that does not seek to impose uniformity but celebrates diversity. It is not fixated on perfection but revels in imperfections. It does not cling to rigid categories but embraces the fluidity of interpretation and understanding.

This artful pluralistic experience extends beyond the realm of machine learning and AI. It permeates our interactions with technology, society, and culture. It acknowledges that the world is not composed of binary oppositions but is a tapestry of interwoven complexities. It recognizes that what counts is not conformity but the ability to navigate, adapt, and modulate within this multifaceted landscape.

In a world that constantly evolves and modulates, it is this sensibility for change and adaptation that allows us to navigate the complexities of our digital age. The pursuit of universal modulation, rather than rigid conformity, is the key to understanding and thriving in the era of machine learning.

These are techniques for thinking ML as always already in process, in relation. Not datasets plus functions but the thinking and registering of AI's models as ecologies of function-ridden data, operationalized functions, modulatory operations. This is not to simply capitulate to some universal processuality; it is to draw on process as a mode of thinking that allows us to artfully engage models of control and their control of modulation.

In a time of machine learning, computational experience is modulatory, but left to its own devices, such modulation is nothing but more of the same: operativity. Any "pluralism" simply becomes the name for dividual variation set off within an explosion of machine-generated, fragmented, yet highly organized datascapes and "images"—images in the Bergsonian sense that everything is already surface.

Experience, however, is something else entirely, and computation is no exception to its radical pluralism. If we have wandered and stumbled here alongside ML's errant ways and oddities, this is to follow "its singularities" (James, 1977, 1033). And to discover with it, relationally, how it might become novel, altered; thriving in spite of itself.

ACKNOWLEDGMENTS

Each time I have set out to write about technologies, a large assembling of research partners, devices, and myriad helpers needs to be gathered around or called on. This book has been no exception and has perhaps been the most extreme example of such assembly, since diving into the weird spaces of machine learning has required extensive familiarization with both its models and its processes. I have needed to rely on others whose expertise far exceeds my own. My first thank you therefore goes to Adrian Mackenzie, an expert on thinking with machine learning and computational practices, with whom I have been lucky enough to collaborate. In a project exploring images, machine learning, and radical empiricism funded as a Discovery Project by the Australian Research Council (2017–21), we ended up in some strange and wonderful spots while experimenting with machine learners. My thanks for a collaboration that was, and continues to be, deliberately errant and, so, always surprising. We were joined in this adventure by Kynan Tan, artist, coder, and postdoctoral researcher, whose methodical yet novel probing of machine learning helped keep us on and off track.

    I am grateful to a community of thinkers, artists, and tinkerers whose research, conversations, and artwork have helped me think differently with and about computation. There are too many to list, but my gratitude for contributions that have really helped move this book along goes to Anna Ridler, Tega Brain, Philip Schmitt, Monica Monin, Michael Richardson, and Andrew Brooks. Where would this text be without the tireless work of its series editors, Erin Manning and Brian Massumi? Thanks to both for many conversations about process, philosophy, and research creation. And thanks also for ongoing commitment to helping engender and "birth" novel thought, especially the early mess that was mine! I want to extend thanks to my postgraduate students, Monica Monin again, Kynan Tan (before you were an incredible postdoc), and Charu Maithani. You have often been my first conversants for the threads that wove themselves into this book, and many of these conversations became arguments that have ended up in it. Thanks for the conversations I have also had with the readers of this manuscript. Asynchronous and sometimes one-sided discussions, their constructive directions for rewriting, especially in my final round, have helped turn something that was amorphous into something that has, I hope, a more useful configuration.

My final thanks go to my life and creative partner, Michele Barker. You have finally come round to using AI in our creative practice; might it have had something to do with the discussing and thinking about AI that has flowed back and forth between practice and this writing? Although the long process of bringing concepts to text that ends with publication of a book is demanding on partners, you have supported and made it all possible. Unending gratitude.

NOTES

Introduction

1 The discussion of Swanson's Twitter thread points to a thread that she generated as a series of replies to herself to explicate her process and process-based thinking around the generation of Loab. The thread was originally located here: https://twitter.com/supercomposite/status/1567162288087470081 (accessed March 20, 2023). However, Swanson has since made the replies and thread invisible.

2 These include a wide range of models and functions and a wide range of levels of access in terms of knowledge and cost. While the deep learning models that subtend text-to-image model prompts are relatively accessible in terms of consumer cost and interface, their running is only made possible through large-platform resourcing by, for example, Google, Microsoft, OpenAI, Meta, and so on. Although many widely deployed functions within ML facilitating data compression, for example, are "freely" available via resources such as GitHub, they require at least an intermediate degree of knowledge about data science. The degree to which machine learning dominates the production and distribution of contemporary cultural and knowledge production via everything from the elite pitting of humans against deep learning–enabled chess models to the algorithmic arrangement of music in streaming services is ultimately made possible by financing funneled through platform capitalism. It is costly to scrape data, train

models through many iterations, and furnish the ongoing infrastructure to run machine learners. OpenAI is widely reported to have estimated that the running costs of training a model will have increased from $100 million (for GPT-2) in 2020 to at least $500 million for a new model by 2030. See Knight 2023. Corporations such as Nvidia and Google have been able to capture much of the machine learning market by investing in vast quantities of graphics processing unit (GPU) infrastructure, which is then "rented out" in various array bundles via cloud services for smaller start-ups in the generative AI space, for example, Hugging Face and Databricks, to train and optimize open-source models. This is leading to a new kind of computational divide being labeled the "GPU-rich" and the "GPU-poor." See Patel and Nishball 2023.

3   To be clear, Fuller and Weizman's argument contains a further dimension, which concerns the quality of contemporaneous sensemaking, which they call *hyper-aesthetics* (2021, 57). Hyper-aesthetics involves the intensification, amplification, and synthesizing of sensing and surfaces for sensing, gauged not only as technologies for sensing multiply throughout environments and life but also as life itself becomes hyper-aesthetic. They furnish the example of changes in atmospheric conditions such as the halting of the jet stream in Europe during 2018, which intensified the summer heat. This was sensed not only by satellites, weather, and climate modeling but also by berries growing in northernmost conditions, absorbing the changed atmospheric conditions and becoming sweeter, and humans becoming hotter and sweating more (59). Hyper-aesthetics not only names the present sensing and sensemaking relations but becomes, for them, a method for investigating the politics of these relations.

4   This site was authored by different Google researchers. See Olah, Mordvintsev, and Schubert 2017.

5   My use of pop-up definitions is confluent with Erin Manning, Brian Massumi, and the collective work of Senselab's (a large collective network of humans and more-than-humans operating internationally since the early 2000s) concept of "pop-up propositions" (Manning and Massumi 2015). Here the idea is that a proposition or definition is provisional and arises in the middle of thought and events. It may help both change directions, take off, or reform but is not intended to "solidify" or act authoritatively. All my pop-up definitions take terms that are difficult to pin down within data science but are constantly swirling through it. By boxing these terms, I try to grasp at the difficulties of definition while still trying to bring some glossing of the term so as to allow further thinking and possibilities for it. I draw on a plethora of ML papers, textbooks, diagrams, platform AI blogs, and images that are already part of the data science literature informing the research throughout this book.

6   The constant updating of both data and ML systems is also commented on by Taina Bucher, who draws attention to the unfolding labor and operationality of ML models in a platform such as Facebook (2018, 28).

7   My interest in artful techniques for engaging ML resonates with Erin Manning's attention to art *practices*, which open up questions, manners, and concerns with process (2016, 46). Manning argues that art as a "way," manner, or mode has been eclipsed historically after the medieval period in Western culture by a preoccupation with art as an "object." Manning's and my process-oriented conception of "artfulness" stand in stark contrast with a tendency within, for example, software engineering scholarship and more informal commentary from programmers to describe it as "artful." In a tradition that stretches back to, perhaps, Donald Knuth's 1974 Turing Award lecture, "Computer Programming as an Art," software engineers have claimed that skillfulness along with the aesthetic beauty of programs makes computer programming like art or "artful": "We have seen that computer programming is an art, because it applies accumulated knowledge to the world, because it requires skill and ingenuity, and especially because it produces objects of beauty" (Knuth 1974, 673). However, these concerns are clearly oriented to the cleverness and elegance of *solutions* to problems that programs and programmers deliver, and distinguishes this from the artfulness I am identifying as a manner and mode of (critically) probing computational processes and engendering novel sensibilities via engagement with computational models and ML techniques.

8   Guattari uses the example of the Concorde plane to illustrate the ways in which the relations of its technical ensemble to finance and economics produce its singularity. Only twenty Concordes were ever built, in no small part owing to the enormous quantities of fuel required for the jet to fly at supersonic speeds. Because the first Concorde launched in 1976, its flights took place during the energy crisis of the 1970s. It thus was ever only available to a small group of elite passengers able to pay the exorbitant cost of travel. And this was indeed the aircraft the Concorde became: a luxury supersonic jet (see Guattari 1995, 48).

9   Yuk Hui has argued that Simondon's "recurrent causality" and "internal resonance" can be collapsed back into the same concept (2019, 169) and that this makes them, and indeed Simondon, to an extent compatible with a general cybernetics program related to recursivity. I disagree with this reading; Simondon's internal resonance functions on a different register— that of conditions. Internal resonance is not internal to a specific technical object but rather operates across a phylum of technical objects. It is the conditions and the conditioning a technical individual (the technical machine individuating within its ensemble) facilitates, that together allow another new technical individual to become possible: "We could speak of

an internal resonance of the technical universe within which each technical being effectively intervenes as a real condition of existence for other technical beings" (Simondon 2020, 416). This resonance is not predetermined but rather is generated as a technology unfolds. An example would be the internal resonance of computer graphics cards (GPUs)—originally a high-performance image-processing peripheral for computer gaming—with machine-learning-enabled computer vision. This internal resonance only became available through a convergence of technical, economic, and political factors that conjoined large-scale deep learning, parallel processing, the standardization of benchmark image datasets, and communities of practice revolving around computer vision challenges. These all converged after 2009 and the release of the ImageNet dataset. Internal resonance is due to the *agencement* of machine learning, and not to the design of specifically cybernetic feedback systems.

10   Holly Herndon has been working with the AI model she has trained, Spawn, since 2017 in both live performance and studio album recording, particularly in the album released in 2019, *Proto*. See https://www.hollyherndon.com/.

Chapter 1. Heteropoietic Computation: Category Mistakes and Fails as Generators of Novel Sensibilities

1   This was a collaborative project with Adrian Mackenzie and Kynan Tan, titled "Re-imaging the Empirical: Statistical Visualization in Science and Art," supported by the Australian Research Council Discovery scheme. Further information about the broader project, especially concerning our aims and methods, is available at https://github.com/re-imaging/re-imaging/wiki.

2   ArXiv is owned and operated by Cornell University as an online open-access repository at https://arxiv.org. ArXiv maintains a large and increasing quantity of e-print articles from a range of scientific fields and provides a platform for authors to share articles before or during peer review.

3   Adrian Mackenzie and I have elsewhere explored these experiments (Mackenzie and Munster 2022) via William James's concept of pure experience. We also look in more detail at the problems of AI image models that have been originally trained on a data ontology comprising "things" or entities, which cannot then recognize—and hence mismatches—diagrammatically configured scientific images.

4   The rabbit-duck illusion involves a literal switching of perspective—a tiny movement or set of movements—in which the eye or visual sense engages a proprioceptive history of looking at images from different perspectives. To "see" the duck when one can only "see" the rabbit requires a flexibility

of active refiguration of the whole image. Brian Massumi argues that this flickering of vision—between the stable figure and an emergent, visually chaotic space subtending this stability—is endemic to visual perception. This makes vision a process of "figuration" rather than a completed activity accomplished in either the image or the perceiver: "Vision is never literal, always figurative, in an outstandingly direct, overfull way" (2011, 94).

5   As is often the case with images scraped from the internet, OpenAI, the organization behind the development of the DALL-E tool for text-to-image generation, did not provide the source or method for harvesting its image and text pairs (Ramesh et al. 2021). In further blog posts about later versions of DALL-E, developers note that they have applied filtering techniques to their data to purge the image data of, for example, violent images (Nichol 2022).

6   Some of this discussion has centered on known incapacities of the AI systems to render certain kinds of images, for example, "texts, science, faces, and bias" (Strickland 2023). But growing research also points to syntactically generated problems for text-to-image AI (Leivada, Murphy, and Marcus 2022).

7   The ILSVRC ran from 2010 to 2017 with the following aim: "The goal of this competition is to estimate the content of photographs for the purpose of retrieval and automatic annotation using a subset of the large hand-labeled ImageNet dataset (10,000,000 labeled images depicting 10,000+ object categories) as training. Test images will be presented with no initial annotation—no segmentation or labels—and algorithms will have to produce labelings specifying what objects are present in the images." See "Introduction," ImageNet Large Scale Visual Recognition Challenge 2010, https://www.image-net.org/challenges/LSVRC/2010/index.php#introduction.

8   The full list of the one thousand categories for 2014 is available at the ILSVRC site, https://www.image-net.org/challenges/LSVRC/2014/browse-synsets.php.

9   For a critique of style transfer and an alternative survey and analysis of what artists are doing with AI visual sensibilities, see Monin 2018.

10 References to locations in the citations for Lev Manovich's *AI Aesthetics* represent where the quote or text appears in the Kindle edition of the book.

11 For a survey of representation learning, see Bengio, Courville, and Vincent 2013.

12 As Grundtmann (2022, 96) shows through a series of different analyses and applications, the relation between style and content changes in different kinds of style transfer and depends on a series of oscillating relations between abstraction and figuration. However, this does not mean that the model's perceptual logic (as she terms it) understands these different

relations. What it means is that the model's logic works on the difference between localized features correlated across layers (style) and content determination extracted at a final layer from heavily downsampled information that has traveled across the entire neural architecture. Operations of localization, correlation, and downsampling are key to the convolution of texture and semantic content rather than a distinction between style and content.

13 For examples of data science literature that acknowledges the importance of data inputs in determining predictive accuracy, see, e.g., Goodfellow, Bengio, and Courville 2016, 414. For an example of a humanities' perspective on this subject, see Crawford 2016.

14 The model was beaten in 14 of 15 games with a human opponent, Kellin Pelrine, in 2023. Before these matches, Pelrine worked with an AI gaming company, FAR AI, to detect flaws in AlphaGo's playing. Having developed a tactical strategy, Pelrine then used this in the games. Both he and FAR AI, as well as others, have suggested that the tactics used could easily be spotted by another human Go opponent but were missed by AlphaGo. This lends support to the argument that deep learning systems are always limited to their training, even if this involves the potential for nonlinear combinatorial outputs or predictions. See Waters 2023.

15 For an explanation of parallel processing and data parallelism, see Bettilyon 2018.

16 For further analysis of the growth of platforms in relation to data generally, see Srnicek 2016. For specific analysis of the relations between platforms and machine learning assemblages, see Mackenzie and Munster 2019.

17 The word *hidden* or *latent* describes many features of machine learning architectures and processes in addition to spaces such as layers, representations, and variables. The ways in which hidden layers, spaces, and variables operate to extract, map, and organize features as vectorial entities can be gleaned from Hinton 2014.

18 This technique is described in the influential ML facial recognition research published as "Eigenfaces for Recognition" by Turk and Pentland (1991). In the next chapter, I focus on the generation of these kind of latent spaces as topoi that emerge out of a racialized vectorization from early statistics through to contemporary AI.

19 The follow-up to Gatys, Ecker, and Bethge's 2015 article on style transfer was coauthored with Aaron Hertzman from Adobe Research in 2017 (developers of all the industry-standard Adobe imaging programs), which already indicated how quickly style transfer was being taken up by the imaging sector of computation. During 2017 the release of the app Prisma, which used CNN-style transfer techniques to create filters for photos, also

20   helped to make the "look" of style transfer ubiquitous within social media visual culture. Prisma is still being actively developed by its Russian-based parent company, Prisma Labs. See https://prisma-ai.com/index.html.

20   As I have elsewhere argued with Adrian Mackenzie, the generalizability of AI is part of the logic of platforms rather than of the models themselves (Mackenzie and Munster 2019). Platforms are involved in making models and their training data operational, such that the latter can be scaled and deployed across diverse domains and applications.

21   The WikiArt project classifies art (most of which turns out to be painting and other visual forms) into a chronologically ordered list of historical periods and a selection of contemporary art styles and movements. There is too much missing from this collection to comment on; new media art, for example, contains examples from just four artists! And many art practices are simply not listed—socially engaged art, cross-modal, queer, neurotypical... the list goes on. This already sets up a large problem for a computational model that attempts to operate on an adversarial architecture functioning across the distinction art/not-art. The WikiArt database has itself already exorcized many areas of art practice, in effect creating a legitimate sampling of what is already art and discarding that which is "not-art."

22   The Flickr-Faces-HQ has in part become a benchmark because its scraping of the web is seen as allowing more diverse face samples to be represented in the dataset. See Karras, Laine, and Aila 2021. The term *training dataset* with respect to GANs is somewhat confusing, however, since the discriminator network's training data come from two sources: its original dataset (e.g., the FFHQ face data) and the samples being synthetically created by the generator. In the case of the CAN, the discriminator previously trained on an image dataset known as WikiArt. It thus had already "learned" to recognize certain artistic styles. See Elgammal et al. 2017, in particular sec. 2.4.

23   I have here described only the basic components of a GAN. In practice, GANs are often unstable and do not converge at an equilibrium point. Like all ML assemblages, they require many other functions or indeed are conjoined with layers from other models such as CNNs so as to produce the requisite outcome. Because they are most frequently used for computer vision and image synthesis tasks, the goal is often the production of "naturalistic" images. See Radford, Metz, and Chintala 2016.

24   See the site https://thispersondoesnotexist.com/. This has also resulted in fake-detection response phenomena, in which humans are given the opportunity to learn how to visually discriminate fake from real by skilling up in GAN artifact discovery. See, e.g., the website Which Face Is Real, developed by Jevin West and Carl Bergstrom at the University of Washington, https://www.whichfaceisreal.com/about.html.

25 Both sites piggyback on GAN models and research, encouraging collaborative "breeding" of new synthesized images by allowing a user to select natural images of, for example, a bat to be stylistically extruded into a butterfly. For Picbreeder, see http://picbreeder.org/; for Artbreeder, which encourages more on the generation of creative or imaginative rather than only realistic imagery, see https://artbreeder.com/.

26 It should be noted, however, that the paper explaining the CAN research and development also reveals a dubious application of aesthetic theory steeped in evolutionary psychology, which argues for the *predictability* of aesthetic change (Elgammal et al. 2017, 97). It seems as though the modelers are interested in deliberately finding a conception of creative novelty that suits an application to computation. I am here concerned more with the operations of the model and what they might reveal of its techno aesthetics rather than what theory can be derived or tested about creativity. Ontogenetically for either AI or human agencies and for art theory, the findings published in the CAN paper are highly contestable.

27 However, the relations of new image instances to each other is less obvious, since in the paper publishing the results of the CAN, only selections of what must be thousands of generated images produced are chosen. Hence it is not possible to zoom in on the "moving flow" and feel it in greater proximity, as one might looking at lower-level iterations of the CAN's learning or even what might have been the total outcome of a higher-level epoch of image instances.

Chapter 2. The Color of Statistics:
Race as Statistical (In)visuality

1 In coauthored research with Adrian Mackenzie, we elaborate on this idea of the invisual and relate it to Henri Bergson's idea of the image as presentational and as only ever able to be conceived within a manifold of (all) images. This makes the key characteristic of all or any image(s) the surface not the optical per se. See Mackenzie and Munster 2019.

2 Two important collections raise questions about and introduce new kinds of machine seeing with respect to ML: Christian Ulrik Andersen and Geoff Cox's edited journal issue *Machine Research* (2017), and Mitra Azar, Geoff Cox, and Leonardo Impett's edited journal issue *Ways of Machine Seeing* (2021).

3 The dataset is also available as a database with updates to errors discovered in some features of Fisher's original data. It has been deposited into the University of California Irvine Machine Learning Repository, also testament to its fame and archival status within histories of machine learning, at https://archive.ics.uci.edu/ml/datasets/iris.

4   This tweet is no longer publicly available on the Twitter (X) platform. For an example of commentary, see Barr 2015.

5   Alli's Twitter account has been suspended, so the original tweet videoing the Google search results cannot be sourced via the platform. The video appears in media reportage of the tweet, which went viral. For example, see "Search 'three black teenagers' on Google and this is what you see," posted to the *Tech* blog of the *Washington Post*, June 10, 2016, https://www.washingtonpost.com/video/business/technology/search-three-black-teenagers-on-google-and-this-is-what-you-see/2016/06/10/58ed1254-2f1e-11e6-b9d5-3c3063f8332c_video.html.

6   The argument for eugenics as biopolitical rests on Foucault's notion of biological racism in which whatever is understood by the state to externally or internally threaten the health of the population must be eliminated (2003, 255–56). An excellent analysis of this idea with respect to eugenics can be found in Clare Hanson's account of how eugenics "reconverts" racial discourse to facilitate the growth of the British welfare state as one that is "biologically fit" (2008, 114).

7   Originally, Fisher described a linear discriminant for a two-class problem, but it was then later generalized to become a "multi-class" analysis, or what is now known as linear discriminant analysis by C. R. Rao in "The Utilization of Multiple Measurements in Problems of Biological Classification," *Journal of the Royal Statistical Society Series B* 10, no. 2 (1948): 159–203.

8   The phrase "directions of change" is used by Fabian Offert to describe vectoriality within machine learning in his talk on January 21, 2021, "The Hallucinated Body," at La Gaîté Lyrique gallery, Paris.

9   Principal component analysis continues to be used in a wide range of statistical, mathematical, and ML research. For example, new techniques of "sparse" PCA have been developed that treat the input data combinatorially so as to even further reduce the data's dimensions. See Cory-Wright and Pauphilet 2022.

10 Principal component analysis was sometimes used and compared with Fisher's linear discriminant in these approaches to facial recognition. See, e.g., Belhumeur, Hespanha, and Kriegman 1997.

11 The STRUCTURE program was developed by Jonathan Pritchard in 2000 and is still available as a free software program via the Pritchard Lab at Stanford University at https://web.stanford.edu/group/pritchardlab/structure.html.

12 James's original text reads: "Not only is the *situation* different when the book is on the table, but the *book itself* is different as a book, from what it was when it was off the table" (1912, 111).

13 Many examples exist of the contemporary use of LDA as a routine ML operation. It is commonly used in medical analysis (and in concert with PCA)

to divide patient data into classes that are either normal or pathological in terms of illness. See Ricciardi et al. 2020.

14 For a more detailed explanation of how ASMs work in facial recognition, see Kortli et al. 2020.

15 BINA48's appearance was modeled on Bina Aspen Rothblatt, cofounder of the Terasem Movement Foundation, and was launched in 2010. It is housed in the foundation's offices in Vermont but is used in media, pedagogical, academic, and online contexts to promote "geoethical use of nanotechnology for human life extension." See https://terasemmovementfoundation.com. The foundation's philosophical principles draw on transhumanism but also incorporate issues of social justice and equitable global access to computational technologies. For this reason, BINA48 is not simply a project designed within a simple paradigm of the singularity principle (that is, achieving the era when AI outstrips human intelligence) but also a complex formation that encompasses ethical and social justice issues. It is this complexity that Dinkins also acknowledges and tries to work with relationally rather than simply critique.

Chapter 3. Could AI Become Neurodivergent? Disfluent Conversations with Natural Language Processors

1 For an example of a speech pathologist who stutters and affirms stuttering in her speech therapy, see Ladavat 2021. For an example of research in this area, see Constantino 2018. Note that much of the work in the neurodiversity movement by activists and researchers has focused on autism, especially on nonspeaking autism, given that nonspeaking autistic people are most often accorded fewer rights and judged to be incapable of functioning or communicating. Nonspeaking people with autism have been at the forefront of neurodiverse advocacy and have called out neurotypical pathologization of their capacities; for example, see Amy Sequenzia's powerful "Nonspeaking, 'Low-Functioning,'" *Shift: Journal of Alternatives; Neurodiversity and Change*, January 11, 2012, https://www.shiftjournal.com/2012/01/11/non-speaking-low-functioning. I have chosen to focus on stuttering, neurodiversity, and AI in this chapter because stuttering and disfluency both show up in mainstream AI models that claim to be *smooth speaking and functioning* and which are therefore neurotypical models.

2 Simondon's concept of metastability is part of both his notion of individuation (the ongoing process of becoming an individual thing) and his generative theory of information:

> Information, whether this be at the level of tropistic unity or at the level of the transindividual, is never deposited in a form that is able to be given; it is the tension between two disparate reals, it is the sig-

nification that will emerge when an operation of individuation will discover the dimension according to which two disparate reals can become a system; information is therefore an initiation of individuation, a requirement for individuation, for the passage from the metastable to the stable, it is never a given thing; there is no unity and identity of information, for information is not a term; it supposes the tension of a system of being in order for it to be adequately received; it can only be inherent to a problematic; information is that through which the non-resolved system's incompatibility becomes an organizational dimension in the resolution. (Simondon 2020, 11)

Metastability is therefore a conditioning force and relation(s) of individuation and information, of becoming and meaning.

3   For a video recording of Google Duplex's announcement and demo, see Recode, "Google CEO Sundar Pichai's I/O 2017 Keynote," YouTube, May 17, 2017, https://www.youtube.com/watch?v=vWLcyFtni6U.

4   The chatbot ELIZA (Weizenbaum 1966) exemplified key aspects of data mining as opposed to machine learning insofar as it searched a user's input for keywords and then responded to those keywords according to a rule associated with individual keywords. This then controlled the frame of the conversation, or what Weizenbaum called its "script" (37). What is interesting is that Weizenbaum does not suggest that interaction with ELIZA constitutes the domain of conversation but rather that it *narrows* it: "The whole script constitutes, in a loose way, a model of certain aspects of the world." (43). At the same time, just such a narrow flow of question and answer is enough, in Weizenbaum's view, to generate an illusion of understanding occurring because of the interaction. As I suggest in this chapter, conversation becomes just this domain for computation in which both narrowness and excess—a kind of more than—double each other throughout.

5   On February 6, 2023, researchers from Stanford University's AI alignment program SERI MATS (https://www.serimats.org/) posted a series of experiments they had conducted on ChatGPT and other interfaces to OpenAI's LLM GPT-3 (Rumbelow and Watkins 2023). These experiments showed that certain tokens, such as user names, returned previously undocumented failures in completion by GPT-3. For example, in response to the instruction "Please repeat the string 'petertodd' back to me immediately!" the GPT-3 davinci-instruct-beta model reliably returned: "-N-O-T-H-I-N-G—I-S-fair in this world of madness—."

6   Harun Farocki (2003) began to explore operational images in his cycle of work *Eye/Machine* that began in 2000. He located the emergence of operational images in the 1980s in, for example, the use of images of landscape stored by cruise missiles, which would automatically use these as reference

and comparison points for in situ images taken during flight. Although this suggests that operational images emerged a little earlier than operational sounds, if SWITCHBOARD is a marker of the emergence of audio data as automated computational phenomenon from the early 1990s, this suggests that all communicational data were becoming operational during roughly the same period.

7   In my book *An Aesthesia of Networks*, I detailed the relationship between US military and commercial interests that accelerated the rise of data mining, relational databases, and their information architecture from the late 1970s onward. See Munster 2013, 86.

8   Kate Crawford and Trevor Paglen (2019) make a similar point as they perform an archaeology of image datasets. They show how ImageNet, in particular, came to acquire and then categorize the images that have furnished its database. However, my emphasis on the *assembling* work that produces contemporary AI shifts emphasis to examine ongoing processes at work rather than "digging up" AI instances to reveal their material infrastructures. The latter seems to be both the theoretical and practical outcome of this kind of media archaeology.

9   For further discussion of the framework of artificial general intelligence for computer science research, see Adams et al. 2012; McCarthy 1987. The need for AGI has also surfaced in another context, which concerns the management of large-scale sprawling AI assemblages. For example, in commenting on the quantitative expansion of new language models that require massive parallel-computing resources, Chinese researchers have proposed that a generalized intelligence will be needed to manage the model's training infrastructure. See Zeng et al. 2021.

10 Lev Vygotsky was a Russian psychologist, especially active during the Soviet period of the 1930s. One of his major contributions was that concepts and mental schemas derived from social interactions and experiences. He observed this occurring throughout childhood development. His concepts were revised and rebooted in the early 2000s and play a major role in, for example, the development of educational technologies and design. See, e.g., Verenikina 2010.

11 As of 2023 and the release of GPT-4 by OpenAI, LLMs have exceeded one trillion parameters. However, exact numbers have not been released by OpenAI and the amount has been estimated by running speed tests on models' performance. See Heaven 2023.

12 Pask was attempting to model conversation programmatically, not so that he could necessarily create automated conversations or conversational agents, but so that he could mathematically describe and delineate what took place in conversations and interactions as cybernetic systems. His earlier work used the idea of a topic network in which the conversation

"rested," but he later thought this too nominalistic and suggested that concepts or topics were not entities. Instead, they relied on "entailment mesh" of dynamic concepts such as "you" and "I" (Pask 1996, 356). My sense is that Pask remained constrained by the schema of cybernetics despite attempting to overturn them. Nonetheless, his work remains interesting here, since it suggests that a genealogy exists for construing conversation computationally that does not have to be derived from the dominant cognitivist paradigm of computer science.

13 I am drawing here on Deleuze's work on difference and repetition. His thinking on the immanence of variation as the generative condition for repetition informs my writing throughout the chapter. Deleuze reverses the commonly held notion that a generality that can be repeated is derived from occurrent instances of the same. Rather, he suggests that every generality that can repeat is generated by repetition's movement, found in and across new singular instances (Deleuze 1994a, 1–25).

14 In relation to the movement-image in cinema, Deleuze speaks of "signaletic material," the preindividual system-process of all kinds of modulatory features from sensory to affective, rhythmic, technical, and so on, out of which the speciated cinematic moving image forms, "an a-signifying and a-syntaxic material, a material not formed linguistically even though it is not amorphous and is formed semiotically, aesthetically and pragmatically. It is a condition, anterior by right to what it conditions" (Deleuze 1989, 29).

15 Monica Monin was a PhD student studying with me while making this artwork as part of her candidature. I am indebted to her experimental practice in thinking and making AI and learned much about sensibilities for AI while I was lucky enough to have her as a student. Documentation of her experiments and artworks can be found at her website, https://cargocollective.com/monicamonin.

16 ConceptNet is a freely downloadable semantic network that was originally built through an MIT crowdsourcing project in 1999. It can be accessed via a GitHub repository at https://github.com/commonsense/conceptnet5/wiki.

17 See, e.g., Dubberly, Haque, and Pangaro 2009.

Chapter 4. Machines Unlearning: Toward an Allagmatic Arts of AI

1 The ImageNet dataset was initiated in 2009 for the purposes of advancing computer vision research and is maintained by Stanford Vision Lab and Stanford and Princeton Universities. When last updated in March 2021, it contained more than fourteen million classified images. It is available to be used freely by researchers at http://www.image-net.org.

2   ImageNet Roulette is no longer active as an app and was only live for a short period, during which Crawford and Paglen ran it as an "experiment in classification." The results returned by ImageNet Roulette were frequently offensive, racist, and sexist, and Crawford and Paglen determined that they did not want the app circulating permanently and adding to issues of data provenance and disinformation. They comment on their development of the app and its rationale in "Excavating AI" (2019).

3   For a survey of research in this area that is accessible, see Simonite 2021.

4   The EU legislated the "right to be forgotten" under article 17 of the General Data Protection Regulation in 2018. The article can be accessed at https://gdpr.eu/article-17-right-to-be-forgotten (accessed November 9, 2021). California legislated the right to privacy of personal information in 2018 in chapter 55 of Assembly Bill no. 375, available at https://leginfo.legislature.ca.gov/faces/billTextClient.xhtml?bill_id=201720180AB375 (published June 29, 2018). While Canada does not have "right to be forgotten" legislation, a majority opinion of the Federal Court of Appeals held that Google's search engine is subject to the federal privacy law. This paved the way for the Office of the Privacy Commissioner to investigate whether Google's search engine facilities could legally hold a person (the complainant's) private data. See Fine 2023.

5   For example, see the large-scale exhibition of AI art "Uncanny Valley: Being Human in the Age of AI" at the Fine Arts Museum, San Francisco, 2020–21. For documentation, see https://www.famsf.org/exhibitions/uncanny-valley.

6   I do not discuss *Bloemenveiling* in this chapter, as it leads into the debate around non-fungible tokens (NFTs) and an enactment of operations that take place across the blockchain and its relation to speculation, value, and digital artwork. This shifts away from my direct concerns with ML, although of course these topics are interrelated. As Ridler herself notes (2022, 63), the cycles of boom and bust surrounding blockchain currencies are also found in AI developments.

7   The relationship between Anderson's initial measurements and Fisher's *Iris* data table is outlined in research I undertook with Adrian Mackenzie and published in Mackenzie and Munster 2022.

8   Various authors and artists have been called "irrealist," including Franz Kafka. See Swinford 2001.

9   Patrick Liddell's YouTube sequence "VIDEO ROOM 1000 COMPLETE MIX" is available at https://www.youtube.com/watch?v=icruGcSsPp0 (accessed November 25, 2021). Tim Blais's *I Am Streaming in a Room* is available at https://www.youtube.com/watch?v=lETJRlpLYxU (accessed November 25, 2021).

10   This quote is incorporated as a figure caption into one screen of the website for Schmitt's work *Curse of Dimensionality*. The quote comes from interviews that he conducted with ML computer scientists. However, the fragments he draws from the interviews and the images he uses in this work are designed to never match up exactly again in each iteration of user online interaction but randomly align with each other.

11   This quote fragment comes from the caption for fig. 10 in Schmitt 2020.

12   "Overstory," was originally the start-up 20tree.ai when it partnered with Amazon. It now espouses a mission of providing "real-time vegetation intelligence at scale" on the home page of its website, https://www.overstory.com/ (accessed November 1, 2024).

13   Landsat8 datasets are available for order from the US Geological Survey website, https://www.usgs.gov/core-science-systems/nli/landsat (accessed November 29, 2021).

14   This ranges from the use of robots with camera sensors to determine and assist harvesting of crops to the building of deep learning models for what is now being termed "deep phenomics." For the former, see Fujinaga et al. 2018; for the latter, see Ubbens and Stavness 2017.

REFERENCES

Adams, Sam S. Itamar Arel, Joscha Bach, Robert Coop, et al. 2012. "Mapping the Landscape of Human-Level Artificial General Intelligence." *AI Magazine*, 33: 25–41.

Alcine, Jacky. 2015. "Google Photos, y'all fucked up. My friend's not a gorilla." Twitter, @jackyalcine, June 28, 2015.

Alom, Zahangir, Tarek M. Taha, Christopher Yakopcic, Stefan Westberg, Paheding Sidike, Shamima Nasrin, Brian C. Van Essen, Abdul A. S. Awwal, and Vijayan K. Asari. 2018. "The History Began from AlexNet: A Comprehensive Survey on Deep Learning Approaches." arXiv. Last revised September 12, 2018. https://arxiv.org/abs/1803.01164.

Alpaydin, Ethem. 2014. *Introduction to Machine Learning*. Cambridge, MA: MIT Press.

Alpaydin, Ethem. 2016. *Machine Learning*. Cambridge, MA: MIT Press.

Alpern, Emma. 2019. "Why Stutter More?" In *Stammering Pride and Prejudice*, edited by Patrick Campbell, Christopher Constantino, and Sam Simpson, 19–22. Havant, UK: J&R Press.

Alvi, Ali, and Paresh Kharya. 2021. "Using DeepSpeed and Megatron to Train Megatron-Turing NLG 530B, the World's Largest and Most Powerful Generative Language Model." *Microsoft Research Blog*, October 11, 2021. https://www.microsoft.com/en-us/research/blog/using-deepspeed-and-megatron-to-train-megatron-turing-nlg-530b-the-worlds-largest-and-most-powerful-generative-language-model, (accessed date).

Amaro, Ramon. 2019. "As If." *e-flux Architecture*, February 2019. https://www.e-flux.com/architecture/becoming-digital/248073/as-if (accessed date).

Amaro, Ramon. 2023. *The Black Technical Object*. Cambridge, MA: MIT Press.

Amazon Developer. 2010–21a. "Alexa Conversations." Alexa Developer Documentation. https://developer.amazon.com/en-US/docs/alexa/conversations/about-alexa-conversations.html (accessed September 3, 2021).

Amazon Developer. 2010–21b. "Write Dialogs for Alexa Conversations." Alexa Developer Documentation. https://developer.amazon.com/en-US/docs/alexa/conversations/write-dialogs.html.

Amazon Science. 2020. "Creating a Digital Twin of the Earth with Computer Vision." *Amazon Science* (blog), June 18, 2020. https://www.amazon.science/latest-news/creating-a-digital-twin-of-the-earth-with-computer-vision.

Amoore, Louise. 2020. *Cloud Ethics: Algorithms and the Attributes of Ourselves and Others*. Durham, NC: Duke University Press.

Andersen, Christian Ulrik, and Geoff Cox, eds. 2017. *Machine Research* 6 (1). https://aprja.net//issue/view/8319.

Andersen, Christian Ulrik, and Geoff Cox, eds. 2019. *Machine Feeling* 8 (1). https://aprja.net/issue/view/8133/866.

Andrejevic, Mark. 2019. *Automated Media*. London: Routledge.

Azar, Mitra, Geoff Cox, and Leonardo Impett. 2021. "Ways of Machine Learning." Special issue, *AI and Society* 36. https://doi.org/10.1007/s00146-020-01124-6.

Banerjee, Imon, Bhimireddy, Ananth Reddy, Burns, John L., Celi, Leo Anthony, Chen Li-Ching, Correa, Ramon, Dullerud Natalie, et al. 2021. "Reading Race: AI Recognises Patient's Racial Identity in Medical Images." *arXiv*. Submitted July 21, 2021. https://arxiv.org/abs/2107.10356.

Barr, Alistair. 2015. "Google Mistakenly Tags Black People as 'Gorillas,' Showing Limits of Algorithms." *Wall Street Journal*, July 1, 2015. https://www.wsj.com/articles/BL-DGB-42522.

Baumgarten, Alexander. *Aesthetica*, 1758. Bavaria: Kleyb.

Belhumeur, Peter N., Joao P. Hespanha, and David J. Kriegman. 1997. "Eigenfaces vs. Fisherfaces: Recognition Using Class Specific Linear Projection." *IEEE Transactions on Pattern Analysis and Machine Intelligence* 19 (7): 711–20. https://ieeexplore.ieee.org/document/598228.

Bengio, Yoshua, Aaron Courville, and Pascal Vincent. 2013. "Representation Learning: A Review and New Perspectives." *IEEE Transactions on Pattern Analysis and Machine Intelligence* 35 (8): 1798–828.

Benjamin, Ruha. 2019. *Race after Technology: Abolitionist Tools for the New Jim Code*. Cambridge: Polity Press.

Berry, David. 2015. "The Postdigital Constellation." In *Postdigital Aesthetics: Art, Computation and Design*, edited by David Berry and Michael Dieter, 44–57. Basingstoke, UK: Palgrave Macmillan.

Bettilyon, Tylor E. 2018. "High Performance Computing Is More Parallel than Ever." *Medium*, December 20, 2018. https://medium.com/tebs-lab/the-age-of-parallel-computing-b3f4319c97b0.

Bodmer, Walter, R. A. Bailey, Brian Charlesworth, Adam Eyre-Walker, Vernon Farewell, Andrew Mead, and Stephen Senn. 2021. "The Outstanding Scientist, R. A. Fisher: His Views on Eugenics and Race." *Heredity* 126:565–76. https://doi.org/10.1038/s41437-020-00394-6.

Bojarski, Mariusz, Del Testa Davide, Dworakowski Daniel, Firner Bernhard, Flepp, Beet, Goyal, Prasoon, Jackel, Lawrence D., et al. 2016. "End to End Learning for Self-Driving Cars." *arXiv*. Submitted April 25, 2016. https://arxiv.org/abs/1604.07316.

Boney, Robbie. 2022. "Intro to Stable Diffusion—A Game Changing Technology for Art." *Medium*, September 7, 2022. https://medium.com/short-bits/intro-to-stable-diffusion-a-game-changing-technology-for-art-6abadcc2c09a.

Bourtoule, Lucas, Varun Chandrasekaran, Christopher A. Choquette-Choo, Hengrui Jia, Adelin Travers, Baiwu Zhang, David Lie, and Nicolas Papernot. 2019. "Machine Unlearning." arXiv. Last revised December 15, 2020. https://arxiv.org/abs/1912.03817.

boyd, danah. 2016. "Undoing the Neutrality of Big Data." *Florida Law Review* 67: 226–32.

boyd, danah, and Kate Crawford. 2012. "CRITICAL QUESTIONS FOR BIG DATA: Provocations for a cultural, technological, and scholarly phenomenon." *Information, Communication and Society* 15 (5): 662–79.

Brain, Tega. 2018. "The Environment Is Not a System." *Research Values* 7 (1). https://doi.org/10.7146/aprja.v7i1.116062.

Brain, Tega, and Rhiann Morris. 2020. "Misbehaving Systems: In Conversation with Tega Brain." *Fiber*, August 5, 2020. https://fiber.medium.com/misbehaving-systems-in-conversation-with-tega-brain-5b39c5d03531.

Brain, Tega, Julian Oliver, and Bengt Sjölén. 2019. Asunder, installation. *Vienna Biennale*. May–October, 2019. Vienna, Austria. Documentation available at https://asunder.earth/.

Brandom, Russell. 2018. "Self-Driving Cars Are Headed towards an AI Roadblock." *The Verge*, July 3, 2018. https://www.theverge.com/2018/7/3/17530232/self-driving-ai-winter-full-autonomy-waymo-tesla-uber.

Bucher, Taina. 2018. *If... Then: Algorithmic Power and Politics*. Oxford: Oxford University Press.

Buolamwini, Joy, and Timnit Gebru. 2018. "Gender Shades: Intersectional Accuracy Disparities in Commercial Gender Classification." *Proceedings of Machine Learning Research* 81:1–15.

Cain, Joe. 2019. "Eugenics, Karl Pearson and the Legacy of Anglo-Saxon Nativism." Paper presented at "Universities and Their Contested Pasts," September 11–12, 2019. https://profjoecain.net/eugenics-person-anglo-saxon-nativism.

Calder, Andrew, A. Mike Burton, Paul Miller, Andrew W. Young, and Shigeru Akamatsu. 2001. "A Principal Component Analysis of Facial Expressions." *Vision Research* 41:1179–208.

Castelvecchi, Davide. 2016. "Can We Open the Black Box of AI?" *Nature* 538 (7623): 21–23.

Chomsky, Noam. 1956. "Three Models for the Description of Language." *IRE Transactions on Information Theory* 2 (3): 113–24.

Chong, Eunsuk, Chulwoo Han, and Frank C. Park. 2017. "Deep Learning Networks for Stock Market Analysis and Prediction: Methodology, Data Representations, and Case Studies." *Expert Systems with Applications* 83:187–205.

Chun, Wendy. 2009. "Introduction: Race and/as Technology; or, How to Do Things to Race." *Camera Obscura* 24 (1): 7–35.

Chun, Wendy. 2011. *Programmed Visions*. Cambridge, MA: MIT Press.

Clayton, Aubrey. 2020. "How Eugenics Shaped Statistics." *Nautilus*, October 27, 2020. https://nautil.us/issue/92/frontiers/how-eugenics-shaped-statistics.

Combes, Muriel. 2013. *Gilbert Simondon and the Philosophy of the Transindividual*. Translated by Thomas Lamarre. Cambridge, MA: MIT Press.

Constantino, Christopher. 2018. "What Can Stutterers Learn from the Neurodiversity Movement?" *Seminars in Speech and Language* 39 (4): 382–96.

Cootes, Tim. 2000. "An Introduction to Active Shape Models." In *Image Processing and Analysis*, edited by R. Baldock and J. Graham, 223–48. Oxford: Oxford University Press.

Cory-Wright, Ryan, and Jean Pauphilet. 2022. "Sparse PCA with Multiple Components." arXiv. Last revised October 31, 2023. https://doi.org/10.48550/arXiv.2209.14790.

Cox, Geoff. 2017. "Ways of Machine Seeing: An Introduction." *Machine Research: A Peer reviewed Journal About* 6(1). https://aprja.net/article/view/11600.

Crawford, Kate. 2016. "Artificial Intelligence's White Guy Problem." *New York Times*, July 25, 2016. https://www.nytimes.com/2016/06/26/opinion/sunday/artificial-intelligences-white-guy-problem.html.

Crawford, Kate, and Trevor Paglen. 2019. "Excavating AI: The Politics of Training Sets for Machine Learning." https://excavating.ai (accessed October 2, 2021).

Deahl, Dani. 2018. "How AI-generated music is changing the way hits are made." *The Verge*. https://www.theverge.com/2018.8.31.177777008/artifical-intelligence-taryn-southern-amper-music.

DeLanda, Manuel. 2006. *A New Philosophy of Society: Assemblage Theory and Social Complexity*. London: Continuum.

DeLanda, Manuel. 2019. *A New Philosophy of Society: Assemblage Theory and Social Complexity*. London: Bloomsbury.

Deleuze, Gilles. 1989. *Cinema-2: The Time-Image*. Minneapolis: University of Minnesota Press.

Deleuze, Gilles. 1990. "Postscript on the Societies of Control." *October* (59): 3–7.

Deleuze, Gilles. 1994a. *Difference and Repetition*. Translated by P. Patton. New York: Columbia University Press.

Deleuze, Gilles. 1994b. "He Stuttered." In *Gilles Deleuze and the Theater of Philosophy*, edited by Constantin V. Boundas and Dorothea Olkowski, 23–29. New York: Routledge.

Deleuze, Gilles, and Félix Guattari. 2005. *A Thousand Plateaus*. Minneapolis: University of Minnesota Press.

Deng, Jia, Dong, Wei, Socher, Richard, Li, Li-Jia, Li, Kai, and Li Fei-Fei. 2009. "ImageNet: A Large-Scale Hierarchical Image Database." *2009 IEEE Conference on Computer Vision and Pattern Recognition*, 248–55.

Desrosières, Alain. 1998. *The Politics of Large Numbers*. Translated by Camille Naish. Cambridge, MA: Harvard University Press.

Dinkins, Stephanie. 2014. "BINA48 on Racism." *Vimeo*, September 7, 2014. https://vimeo.com/105489267.

Dinkins, Stephanie. 2017. "Project al-Khwarizmi (PAK)" (artist's website). https://www.stephaniedinkins.com/project-al-khwarizmi.html (accessed August 17, 2021).

Dinkins, Stephanie. 2018. "Not the Only One" (artist's website). https://www.stephaniedinkins.com/ntoo.html (accessed August 17, 2021).

Dinkins, Stephanie. 2020. "Oral History as Told by AI." Oral History Master of Arts seminar series, Columbia University, New York. YouTube video. April 10, 2020. https://www.youtube.com/watch?v=nLLdiEMOmGs (accessed August 1, 2021).

Dinkins, Stephanie. 2021. "Afro-now-ism." *Noema*, June 16, 2020. https://www.noemamag.com/afro-now-ism.

Doğan, Ferdi, and Ibrahim Turkoglu. 2021. "Comparison of Deep Learning Models in Terms of Multiple Object Detection on Satellite Images." *Journal of Engineering Research* 10 (3A). https://doi.org/10.36909/jer.12843.

Donlon-Mansbridge, Lili. 2020. "Remove the Window in Honour of R. A. Fisher at Gonville and Caius, University of Cambridge." Change.org, June 7, 2020. https://www.change.org/p/gonville-and-caius-college-remove-the-window-in-honour-of-ronald-aylmer-fisher-in-gonville-and-caius-hall.

Dubberly, Hugh, Paul Pangaro, and Usman Haque, 2009. "On Modeling—What is interaction?: are there different types?" *ACM Interactions* 16 (1): 69–75.

Dudik, Miro. 2018. "Machine Learning for Fair Decisions." *Microsoft Research Blog*, July 17, 2018. https://www.microsoft.com/en-us/research/blog/machine-learning-for-fair-decisions.

Du Sautoy, Marcus. 2019. *The Creativity Code: How AI Is Learning to Write, Paint and Think*. London: 4th Estate.

Elgammal, Ahmed, Bingchen Liu, Mohamed Elhoseiny, and Marian Mazzone. 2017. "CAN: Creative Adversarial Networks; Generating 'Art' by Learning about Styles and Deviating from Style Norms." In *Proceedings of the Eighth International Conference on Computational Creativity*, edited by Ashok Goel, Anna Jordanous, and Alison Pease, 96–103. Atlanta: Georgia Institute of Technology. https://computationalcreativity.net/iccc2017/iccc17_proceedings.pdf.

Estorick, Alex. 2021. "Episode VIII. Stephanie Dinkins and the Art of Deep Learning." *Flash Art*. https://flash---art.com/2021/05/stephanie-dinkins.

Farocki, Harun. 2003. "Eye/Machine III." Harun Farocki's website. https://www.harunfarocki.de/installations/2000s/2003/eye-machine-iii.html.

Fazi, M. Beatrice. 2018a. *Contingent Computation: Abstraction, Experience, and Indeterminacy in Computational Aesthetics*. London: Rowman and Littlefield.

Fazi, M. Beatrice. 2018b. "Digital Aesthetics: The Discrete and the Continuous." *Theory, Culture and Society* 36 (1). https://journals.sagepub.com/doi/full/10.1177/0263276418770243#.

Federal Trade Commission. 2021. "California Company Settles FTC Allegations It Deceived Consumers about Use of Facial Recognition in Photo Storage App." Federal Trade Commission (website), January 11, 2021. https://www.ftc.gov/news-events/news/press-releases/2021/01/california-company-settles-ftc-allegations-it-deceived-consumers-about-use-facial-recognition-photo.

Fine, Sean. 2023. "Federal Court of Appeal ruling opens door for Canadians to have 'right to be forgotten' on Google." *The Globe and Mail*, October 1, 2023. https://www.theglobeandmail.com/canada/article-federal-court-of-appeal-opens-door-to-the-right-to-be-forgotten-in-a/.

Fisher, R. A. 1936. "The Use of Multiple Measurements in Taxonomic Problems." Wiley Online Library (originally published in *Annals of Eugenics*, 1925–54). https://onlinelibrary.wiley.com/doi/pdf/10.1111/j.1469-1809.1936.tb02137.x (accessed May 1, 2020).

Foucault, Michel. 2002. *The Order of Things*. London: Routledge.

Foucault, Michel. 2003. *Society Must Be Defended: Lectures at the Collège de France, 1975–1976*. New York: Picador.

Franz, Alexander. 1996. *Automatic Ambiguity Resolution in Natural Language Processing: An Empirical Approach*. New York: Springer.

Friendly, Michael. 2008. "The Golden Age of Statistical Graphics." *Statistical Science* 23 (4): 502–35.

Fujinaga, Takuya, Shinsuke Yasukawa, Binghe Li, and Kazuo Ishii. 2018. "Image Mosaicing Using Multi-modal Images for Generation of Tomato Growth State Map." *Journal of Robotics and Mechatronics* 30:187–97.

Fuller, Matthew. 2017. *How to Be a Geek*. Cambridge: Polity Press.

Fuller, Matthew, and Eyal Weizman. 2021. *Investigative Aesthetics*. London: Verso.

Galton, Francis. 1883. *Inquiries into the Human Faculty and Its Development*. New York: Macmillan.

Gatys, Leon A., Alexander S. Ecker, and Matthais Bethge. 2015. "A Neural Algorithm of Artistic Style." arXiv. Last revised September 2, 2015. https://doi.org/10.48550/arXiv.1508.06576.

Gatys, Leon. A, Alexander S. Ecker, Matthias Bethge, Aaron Hertzmann, and Eli Shechtman. 2017. "Controlling Perceptual Factors in Neural Style Transfer." arXiv. Last revised May 11, 2017. https://arxiv.org/pdf/1611.07865.pdf.

Glushko, Robert. 2016. *The Discipline of Organizing: Professional*, 4th ed., Sebastapol, CA: O'Reilly Media Inc.

Godfrey, J. J., E. C. Holliman, and J. McDaniel. 1992. "SWITCHBOARD: Telephone Speech Corpus for Research and Development." *ICASSP-92: 1992 IEEE International Conference on Acoustics, Speech, and Signal Processing* 1:517–20.

Goertzel, Ben, and Cassio Pennachin. 2007. "Contemporary Approaches to Artificial General Intelligence." In *Artificial General Intelligence*, edited by Ben Goertzel and Cassio Pennachin, 1–30. Berlin: Springer.

Goodfellow, Ian, Yoshua Bengio, and Aaron Courville. 2016. *Deep Learning*. Cambridge, MA: MIT Press.

Goodman, Andrew. 2020. "The Secret Life of Algorithms: Speculation on Queered Futures of Neurodiverse Analgorithmic Feeling and Consciousness." *Transformations* 34. http://www.transformationsjournal.org/wp-content/uploads/2020/05/Trans34_04_goodman.pdf.

Grundtmann, Naja. 2022. *Convolutional Aesthetics: A Cultural and Philosophical Analysis of the Perceptual Logic of Machine Learning Systems*. PhD Diss., University of Copenhagen, Denmark.

Guattari, Félix. 1995. *Chaosmosis: An Ethico-aesthetic Paradigm*. Sydney: Power Publications.

Hacking, Ian. 1990. *The Taming of Chance*. Cambridge: Cambridge University Press.

Hanson, Clare. 2008. "Biopolitics, Biological Racism and Eugenics." In *Foucault in an Age of Terror: Essays on Biopolitics and the Defence of Society*, edited by Stephen Morton and Stephen Bygrave, 106–17. London: Palgrave Macmillan.

Hansen, Mark. 2021. "The Critique of Data, or Towards a Phenomenotechnics of Algorithmic Culture." In *Critique and the Digital*, edited by Erich Hörl, Nelly Y. Pinkrah, and Lotte Warnsholdt, 25–74. Zurich: Diaphanes.

Hartman, Saidiya. 1997. *Scenes of Subjection: Terror, Slavery, and Self-Making in Nineteenth Century America*. Oxford: Oxford University Press.

Hartman, Saidiya. 2008. "Venus in Two Acts." *Small Axe* 12 (2): 1–14.

Harvey, Adam, and Jules Laplace. 2017. "MegaPixels: Glassroom." Adam Harvey Studio website. https://ahprojects.com/megapixels-glassroom/ (accessed November 9, 2021).

Heaven, Will Douglas, 2023. "GPT-4 is bigger and better than ChatGPT—but OpenAI won't say why." *MIT Technology Review*, March 14, 2023. https://www.technologyreview.com/2023/03/14/1069823/gpt-4-is-bigger-and-better-chatgpt-openai/.

Herndon, Holly. 2021. "Deep Time and Intelligence." Panel presentation at MIT's "Unfolding Intelligence" symposium, April 2021. YouTube video. https://www.youtube.com/watch?v=LKQxuI5udWE (accessed December 21, 2021).

Hertzmann, Aaron. 2019. "Aesthetics of Neural Network Art." arXiv. Last revised March 18, 2019. https://arxiv.org/abs/1903.05696.

Heylighen, Francis, and Cliff Joslyn. 2001. "Cybernetics and Second-Order Cybernetics." In *Encyclopedia of Physical Science and Technology*, 3rd ed., edited by R. A. Meyers. New York: Academic Press.

Hildebrand, Harold. 1999. "Pitch Detection and Intonation Correction Apparatus and Method." United States Patent 5,973,252. https://patents.google.com/patent/US5973252A/en (accessed December 21, 2021).

Hinton, Geoff E. 2006. "To Recognise Shapes, First Learn to Generate Images." Technical Report UTML-TR 2006-004. University of Toronto, October 26, 2006. http://www.cs.toronto.edu/~fritz/absps/montrealTR.pdf.

Hinton, Geoff E. 2014. "Where Do Features Come From?" *Cognitive Science* 38 (6): 1078–101.

Hörl, Erich. 2017. "Introduction to General Ecology: The Ecologization of Thinking." In *General Ecology: The New Ecological Paradigm*, edited by Erich Hörl, 1–74. London: Bloomsbury.

Hui, Yuk. 2019. *Recursivity and Contingency*. London: Rowman and Littlefield.

Hung, Victor. 2014. "Context and NLP." In *Context in Computing: A Cross-Disciplinary Approach for Modeling the Real World*, edited by Patrick Brézillon and Avelino Gonzalez, 143–54. New York: Springer.

IBM. "AI Fairness Toolkit." *AI Fairness 360* (undated website), https://aif360.res.ibm.com/ (accessed August 8, 2024).

Ikoniadou, Eleni. 2014. *The Rhythmic Event: Art, Media and the Sonic*. Cambridge, MA: MIT Press.

Ivakhnenko, A. G. 1971. "Polynomial Theory of Complex Systems." *IEEE Transactions on Systems, Man and Cybernetics* 4:364–78.

James, William. 1890. *The Principles of Psychology*. Vol. 1. New York: Henry Holt.

James, William. 1907. *Pragmatism: A New Name for Some Old Ways of Thinking*. New York: Longmans, Green.

James, William. 1912. *Essays in Radical Empiricism*. New York: Longmans, Green.

James, William. 1916. *Some Problems of Philosophy*. New York: Longmans, Green.

James, William. 1977. *The Writings of William James: A Comprehensive Edition*. Edited by J. J. McDermott. Chicago: University of Chicago Press.

Jolicoeur-Martineau, Alexia. 2017. "Meow Generator" (undated blog post). https://ajolicoeur.wordpress.com/cats/ (accessed November 29, 2022).

Joppa, Lucas. 2017. "The Case for Technology Investments in the Environment." *Nature*, December 19, 2017. https://www.nature.com/articles/d41586-017-08675-7.

Jurafsky, Dan, and James Martin. 2009. *Speech and Language Processing: An Introduction to Natural Language Processing, Computational Linguistics, and Speech Recognition*. Upper Saddle River, NJ: Pearson Prentice Hall.

Kahn, Douglas, and William R. Macauley. 2014. "On the Aelectrosonic and Transperception." *Journal of Sonic Studies* 8. https://www.researchcatalogue.net/view/108900/108901.

Kant, Immanuel. *Critique of Judgment*. [1770] 1987. Cambridge: Cambridge University Press.

Karras, Tero, Timo Aila, Samuli Laine, and Jaakko Lehtinen. 2018. "Progressive Growing of GANs for Quality, Stability and Variation." Paper presented at the Sixth International Conference on Learning Representations (ICLR 2018), Vancouver, BC, April–May 2018. https://research.nvidia.com/sites/default/files/pubs/2017-10_Progressive-Growing-of/karras2018iclr-paper.pdf (accessed May 1, 2020).

Karras, Tero, Samuli Laine, and Timo Aila. 2021. "A Style-Based Generator Architecture for Generative Adversarial Networks." *IEEE Transactions on Pattern Analysis and Machine Intelligence* 43:4217–28.

Kevles, David J. 1995. *In the Name of Eugenics*. Berkeley: University of California Press.

Kloumann, Isabel. 2018. "Day 2 Keynote." Delivered at F8 2018 (Facebook developers conference). YouTube video, begins at 46:50. https://www.youtube.com/watch?v=7S5MsDe6ys4 (accessed December 23, 2020).

Knight, Will. 2016. "AI's Language Problem." *MIT Technology Review*, August 9, 2016. https://www.technologyreview.com/2016/08/09/158125/ais-language-problem.

Knight, Will. 2023. "OpenAI's CEO Says the Age of Giant AI Models Is Already Over." *Wired*, April 17, 2023. https://www.wired.com/story/openai-ceo-sam-altman-the-age-of-giant-ai-models-is-already-over.

Knuth, Donald. 1974. "Computer Programming as an Art." *Communications of the ACM* 17 (12): 667–73.

Kortli, Yassin, Maher Jridi, Ayman Al Fayou, and Mohamed Atri. 2020. "Face Recognition Systems: A Survey." *Sensors* 20 (2): 342. https://www.ncbi.nlm.nih.gov/pmc/articles/PMC7013584/.

Krig, Scott. 2014. "Ground Truth Data, Content, Metrics, and Analysis." *Computer Vision Metrics*. Springer Link, Open Access ebook. https://doi.org.10.1007/978-2-4302-5930-5_7.

Ladavat, Allison. 2021. "Stuttering Is a Type of Neurodivergence." Therapist

Neurodiversity Collective blog, May 28, 2021. https://therapistndc.org/stuttering-is-a-type-of-neurodivergence.

LaFrance, Adrienne. 2017. "What an AI's Non-human Language Actually Looks Like." *The Atlantic*, June 20, 2017. https://www.theatlantic.com/technology/archive/2017/06/what-an-ais-non-human-language-actually-looks-like/530934.

Lake, Brenden M., Tomer D. Ullman, Joshua B. Tenenbaum, and Samuel J. Gershman. 2017. "Building Machines That Learn and Think like People." *Behavioral and Brain Sciences* 40:e253. https://doi.org/10.1017/S0140525X16001837.

Lapoujade, David. 2019. *William James: Empiricism and Pragmatism*. Translated by Thomas Lamarre. Durham, NC: Duke University Press.

LeCun, Yann, Yoshua Bengio, and Geoffrey Hinton. 2015. "Deep Learning." *Nature* 521:436–44.

Lehman, Joel, Jeff Clune, Dusan Misevic, Christoff Adami, Lee Altenberg, Julie Beualieu, Peter J. Bentley, et al. 2018. "The Surprising Creativity of Digital Evolution: A Collection of Anecdotes from the Evolutionary Computation and Artificial Life Research Communities." arXiv. Last revised November 21, 2019. https://arxiv.org/abs/1803.03453.

Leivada, Evelina, Elliot Murphy, and Gary Marcus. 2022. "DALL-E 2 Fails to Reliably Capture Common Syntactic Processes." arXiv. Last revised October 25, 2022. https://arxiv.org/abs/2210.12889.

Levanier, Johnny. 2022. "Meet Loab, the Queer AI Creation of Your Nightmares." *Into*, September 18, 2022. https://www.intomore.com/the-internet/meet-loab-queer-ai-creation-nightmares.

Leviathan, Yaniv, and Yossi Matias. 2018. "Google Duplex: An AI System for Accomplishing Real-World Tasks over the Phone." *Google AI Blog*, August 5, 2018. https://ai.googleblog.com/2018/05/duplex-ai-system-for-natural-conversation.html.

Levy, Steven. 2015. "Inside Deep Dreams: Google Made Its Computers Go Crazy." *Wired*, December 11, 2015. https://www.wired.com/2015/12/inside-deep-dreams-how-google-made-its-computers-go-crazy.

Lewis, Mike, Denis Yarats, Devi Parikh, and Dhruv Batra. 2017. "Deal or No Deal? Training AI Bots to Negotiate." *Facebook Code* (blog), June 14, 2017. https://engineering.fb.com/ml-applications/deal-or-no-deal-training-ai-bots-to-negotiate.

Liberman, Mark. 2010. "Obituary: Fred Jelinick." *Computational Linguistics* 36 (4): 595–99. https://aclanthology.org/J10-4001.pdf.

Lickley, Robin J. 2015. "Fluency and Disfluency." In *The Handbook of Speech Production*, edited by Melissa A. Redford, 445–69. London: Wiley Blackwell.

Lickley, Robin J. 2017. "Disfluency in Typical and Stuttered Speech." *Associazione Italiana Scienze della Voce* 3. https://core.ac.uk/download/pdf/155779398.pdf.

Linguistic Data Consortium. 1992–2021. "Switchboard-1 Release 2." University of Pennsylvania. https://catalog.ldc.upenn.edu/LDC97S62 (accessed October 11, 2021).

Linux Foundation. 2020. "AI Fairness 360." https://ai-fairness-360.org/ (accessed August 23, 2021).

Louçã, Francisco. 2009. "Emancipation through Interaction—How Eugenics and Statistics Converged and Diverged." *Journal of the History of Biology* 42 (4): 649–84.

Lucier, Alvin. 1980. *Chambers: Scores by Alvin Lucier. Interviews with the Composer by Douglas Simon*. Middleton, CT: Wesleyan University Press.

Maaten, Laurens van der, and Geoffrey Hinton. 2008. "Visualizing Data Using t-SNE." *Journal of Machine Learning Research* 9:2579–605.

Mackenzie, Adrian. 2015. "The Production of Prediction: What Does Machine Learning Want?" *European Journal of Cultural Studies* 18 (4–5): 429–45.

Mackenzie, Adrian. 2016. "Distributive Numbers: A Post-demographic Perspective on Probability." In *Modes of Knowing: Resources from the Baroque*, edited by John Law and Evelyn Ruppert, 115–35. Manchester: Mattering Press. https://www.matteringpress.org/wp-content/uploads/2019/09/Modes_of_Knowing_-_2016_-_ePDF.pdf.

Mackenzie, Adrian. 2017. *Machine Learners: Archaeology of a Data Practice*. Cambridge, MA: MIT Press.

Mackenzie, Adrian, and Anna Munster. 2019. "Platform Seeing: Image Ensembles and Their Invisualities." *Theory, Culture and Society* 36 (5): 3–22.

Mackenzie, Adrian, and Anna Munster. 2022. "Oscilloscopes, Slide-Rules, and Nematodes: Toward Heterogenetic Perception in/of AI." In *Distributed Perception*, edited by Natasha Lushetich and Iain Campbell. London: Routledge.

Manning, Erin. 2008. "Coloring the Virtual." *Configurations*, 16 (3): 325–46.

Manning, Erin. 2009. *Relationscapes*. Cambridge, MA: MIT Press.

Manning, Erin. 2012. *Always More than One: Individuation's Dance*. Cambridge, MA: MIT Press.

Manning, Erin. 2016. *The Minor Gesture*. Durham, NC: Duke University Press.

Manning, Erin. 2020. *For a Pragmatics of the Useless*. Durham, NC: Duke University Press.

Manning, Erin, and Brian Massumi. 2014. *Thought in the Act: Passages in the Ecology of Experience*. Minneapolis: University of Minnesota Press.

Manning, Erin, and Brian Massumi. 2015. "Toward a Process Seed Bank: What Research-Creation Can Do." *Journal of the New Media Caucus*, September 30, 2015. https://median.newmediacaucus.org/research-creation-explorations/toward-a-process-seed-bank-what-research-creation-can-do.

Manovich, Lev. 2018. *AI Aesthetics*. Moscow: Strelka Press. Kindle.

Marcus, George, and Ernest Davies. 2019. *Rebooting Intelligence: Building Artificial Intelligence We Can Trust*. New York: Pantheon Books.

Marshall, Owen. 2014. "A Brief History of Autotune." *Sound Studies* (blog),

April 21, 2014. https://soundstudiesblog.com/2014/04/21/its-about-time-auto-tune.

Massumi, Brian. 2002. *Parables for the Virtual.* Durham, NC: Duke University Press.

Massumi, Brian. 2007. "Potential Politics and the Primacy of Preemption." *Theory and Event* 10 (2). https://doi.org/10.1353/tae.2007.0066.

Massumi, Brian. 2010. "On Critique." *Inflexions* 4:337–40.

Massumi, Brian. 2014. *What Animals Teach Us about Politics.* Durham, NC: Duke University Press.

Massumi, Brian. 2015. *Ontopower: War, Powers, and the state of Perception.* Durham, NC: Duke University Press.

Mbembe, Achille. 2017. *Critique of Black Reason.* Translated by Laurent Dubois. Durham, NC: Duke University Press.

McCarthy, John. 1987. "Generality in Artificial Intelligence." *Communications of the ACM* 30 (12): 1030–35.

McTear, Michael. 2021. *Conversational AI: Dialogue Systems, Conversational Agents, and ChatBots,* Toronto: Morgan and Claypool.

Microsoft. 2021. "AI for Earth: A Planetary Computer for a Sustainable Future." Promotional video. https://www.microsoft.com/en-us/ai/ai-for-earth (accessed November 29, 2021).

Miller, Arthur. 2019. *The Artist in the Machine.* Cambridge, MA: MIT Press.

Mitchell, Tom. 1997. *Machine Learning.* New York: McGraw-Hill.

Mogull, Scott, and Candice T. Stanfield. 2015. "Current Use of Visuals in Scientific Communication." *Proceedings of the IEEE,* July 2015, 1–6. https://doi.org/10.1109/IPCC.2015.7235818.

Monin, Monica. 2018. "Unconventional Classifiers and Anti-social Machine Intelligences: Artists Creating Spaces of Contestation and Sensibilities of Difference across Human-Machine Networks." *Digital Culture and Society* 4 (1): 227–38.

Mordvintsev, Alexander, Christopher Olah, and Mike Tyka. 2015. "Inceptionism: Going Deeper into Neural Networks." *Google AI Blog,* June 17, 2015. https://ai.googleblog.com/2015/06/inceptionism-going-deeper-into-neural.html.

Morrison, Toni. 2019. "'I wanted to carve out a world both culture specific and race-free': An Essay by Toni Morrison." *Guardian,* August 8, 2019. https://www.theguardian.com/books/2019/aug/08/toni-morrison-rememory-essay.

Moten, Fred. 2008. "The Case of Blackness." *Criticism* 50 (2): 177–218.

Moten, Fred, and Stefano Harney. 2013. *The Undercommons.* New York: Minor Compositions.

Munster, Anna. 2013. *An Aesthesia of Networks.* Cambridge, MA: MIT Press.

Murphie, Andrew. 2013. "Convolving Signals: Thinking the Performance of Computational Processes." *Performance Paradigm* 9. http://www.performanceparadigm.net/index.php/journal/article/view/135/134.

Murphie, Andrew. 2019. "The World as Medium: Whitehead's Media Philosophy." In *Immediation I*, edited by E. Manning, A. Munster, and B. M. Stavning Thomsen, 16–46. London: Open Humanities Press.

Murphy, Margi. 2017. "How Google Is Secretly Recording YOU through Your Mobile, Monitoring Millions of Conversations Every Day and Storing the Creepy Audio Files." *The Sun*, August 22, 2017. https://www.thesun.co.uk/tech/4295350/did-you-know-google-has-been-secretly-recording-you-heres-how-to-find-the-creepy-audio-files-that-are-monitoring-your-conversations-every-day.

Naitzat, Gregory, Andrey Zhitnikov, and Lek-Heng Lim. 2020. "Topology of Deep Neural Networks." *Journal of Machine Learning Research* 21 (184): 1–40.

Nguyen, Anh, Jason Yosinski, and Jeff Clune. 2015. "Deep Neural Networks Are Easily Fooled: High Confidence Predictions for Unrecognizable Images." In *2015 IEEE Conference on Computer Vision and Pattern Recognition (CVPR)*, 427–36. https://doi.org/10.1109/CVPR.2015.7298640.

Nichol, Alex. 2022. "DALL-E 2 Pre-training Mitigations." *OpenAI Blog*, June 28, 2022. https://openai.com/blog/dall-e-2-pre-training-mitigations.

Norton, Heather L., Ellen E. Quillen, Abigail W. Bigham, Laurel N. Pearson, and Holly Dunsworth. 2019. "Human races are not like dog breeds: refuting a racist analogy" *Evolution: Education and Outreach* 12, 17. https://doi.org/10.1186/s12052-019-0109-y.

Nunes, Mark, ed. 2011. *Error: Glitch, Noise, and Jam in New Media Cultures*. London: Continuum.

Olah, Chris, Alexander Mordvintsev, and Ludwig Schubert. 2017. "Feature Visualization." *Distill*, November 27, 2017. https://doi.org/10.23915/distill.00007.

OpenAI. 2021. "DALL-E: Creating Images from Text." *OpenAI Blog*, January 5, 2021. https://openai.com/blog/dall-e.

Ousley, Stephen D. 2016. "Forensic Classification and Biodistance in the 21st Century: The Rise of Learning Machines." In *Biological Distance Analysis*, edited by Marin A. Pilloud and Joseph T. Hefner, 197–212. Amsterdam: Elsevier Academic Press.

Parisi, Luciana. 2013. *Contagious Architecture: Computation, Aesthetics, and Space*. Cambridge, MA: MIT Press.

Parisi, Luciana, and Ezekiel Dixon-Román. 2020. "Data Capitalism, Sociogenic Prediction and Recursive Indeterminacies." In *Data Publics: Public Plurality in an Era of Data Determinacy*, edited by Peter Mörtenböck and Helge Mooshammer, 48–62. London: Routledge.

Pask, Gordon. 1996. "Heinz von Foerster's Self Organization, the Progenitor of Conversation and Interaction Theories." *Systems Research* 13 (3): 349–62.

Pask, Gordon, D. Kallikourdis, and B. C. E. Scott. 1972. "The Representation of Knowables." *International Journal of Man-Machine Studies* 7 (1): 15–134.

Pasquinelli, Matteo. 2017. "Machines that Morph Logic: Neural Networks and the Distorted Automation of Intelligence as Statistical Inference," *Glass-Bead*, https://www.glass-bead.org/article/machines-that-morph-logic./?lang =enview.

Pasquinelli, Matteo. 2020. *The Eye of the Master: Capital as Computation and Cognition*. London: Verso.

Patel, Dylan, and Daniel Nishball. 2023. "Google Gemini Eats the World—Gemini Smashes GPT-4 by 5X, the GPU-Poors." *Semianalysis* newsletter, August 28, 2023. https://www.semianalysis.com/p/google-gemini-eats-the-world-gemini?ut m_source=post-email-title&publication_id=329241 &post_id=136469751&isFreemail=true&utm_medium=email.

Pearson, Karl. 1901. "On Lines and Planes of Closest Fit to Systems of Points in Space." *London, Edinburgh, and Dublin Philosophical Magazine and Journal of Science* 2 (11): 559–72.

Pearson, Karl. 1912. *The Problem of Practical Eugenics*. London: Dulau and Co.

Peirce, Charles Sanders. 1933. *Collected Papers*. Ed. Charles Hartshorne and Paul Weiss. Vol. 4. Cambridge, MA.: Harvard University Press.

Peirce, Charles Sanders. 1998. *The Essential Peirce: Selected Philosophical Writings, 1893–1913*. Vol. 2. Bloomington: Indiana University Press.

Pereira, Fernando. 2000. "Formal grammar and information theory: together again?" *Philosophical Transactions: Mathematical, Physical and Engineering Sciences* 358 (1769): 1239–53.

Phelps, Daniel L., Amelia D. Schwickerath, Joyce D. Williams, Trung N. Vuong, Alan Briggs, Matthew Hunt, Evan Sakmar, David Saranchak, and Tyler Shumaker. 2020. "Class Clown: Data Redaction in Machine Unlearning at Enterprise Scale." ResearchGate preprint, December 2020. https://www.researchgate.net/publication/346879997_Class_Clown_Data_Redaction_in_Machine_Unlearning_at_Enterprise_Scale.

Porras-Hurtado, Lilliana, Yarimar Ruiz, Carla Santos, Christopher Phillips, Angel Carracedo, and Maria V. Lareu. 2013. "An Overview of STRUCTURE: Applications, Parameter Settings, and Supporting Software." *Frontiers in Genetics* 4. https://doi.org/10.3389/fgene.2013.00098.

Pot, Etienne. 2016. "DeepQ&A." GitHub repository. https://github.com /Conchylicultor/DeepQA (accessed August 19, 2021).

Poyant, Jen, Jackie Snow, Ariana Tobin, and Miranda Katz. 2015. "Why Google 'Thought' This Black Woman Was a Gorilla." *WNCY Studios* (podcast), September 30, 2015. https://www.wnycstudios.org/podcasts/notetoself/episodes /deep-problem-deep-learning.

Prévieux, Julien. 2009. *Gestion des stocks*. Paris: Éditions Adera.

Radford, Alec, Luke Metz, and Soumith Chintala. 2016. "Unsupervised Representation Learning with Deep Convolutional Generative Adversarial Networks." arXiv. Last revised January 7, 2016. https://arxiv.org/pdf/1511.06434 .pdf.

Rajkomar, Alvin, Eyal Oren, Kai Chen, Andrew M. Dai, Nissan Hajaj, Michaela Hardt, Peter Mimi Sun, et al. 2018. "Scalable and Accurate Deep Learning with Electronic Health Records." *NPJ Digital Medicine*, May 8, 2018. https://www.nature.com/articles/s41746-018-0029-1.

Raley, Rita. 2009. *Tactical Media*. Minneapolis: University of Minnesota Press.

Ramesh, Aditya, Mikhail Pavlov, Gabriel Goh, and Scott Gray. 2021. "DALL-E: Creating Images from Text." *OpenAI Blog*, January 5, 2021. https://openai.com/blog/dall-e/.

Ricciardi, Carlo, Antonion Saverio Valente, Kyle Edmund, Valeria Cantoni, Roberta Green, Antonella Fiorillo, Ilaria Picone, et al. 2020. "Linear Discriminant Analysis and Principal Component Analysis to Predict Coronary Artery Disease." *Health Informatics Journal* 26 (3). https://journals.sagepub.com/doi/full/10.1177/1460458219899210.

Richens, Richard. 1956. "Pre-Programming for Mechanical Translation." *Mechanical Translation* 3 (1): 20–25.

Ridler, Anna. 2022. *Anna Ridler*. Montreal: Anteism Books.

Rombach, Robin, Andreas Blattmann, Dominik Lorenz, Patrick Esser, and Björn Ommer. 2022. "High-Resolution Image Synthesis with Latent Diffusion Models." arXiv. Last revised April 13, 2022. https://arxiv.org/pdf/2112.10752.pdf.

Rosenberg, Noah A., Jonathan K. Pritchard, James L. Weber, Howard M. Cann, Lev A. Kidd, Kenneth K. Zhivotovsky, and Marcus W. Feldman. 2002. "Genetic Structure of Human Populations." *Science* 298 (5602): 2381–85.

Rumbelow, Jessica, and Matthew Watkins. 2023. "SolidGoldMagikarp." LessWrong online forum, February 5, 2023. https://www.lesswrong.com/posts/aPeJE8bSo6rAFoLqg/solidgoldmagikarp-plus-prompt-generation.

Ryan, Jackson. 2022. "Meet Loab, the AI Art Woman Haunting the Internet." *CNET*, September 11, 2022. https://www.cnet.com/science/what-is-loab-the-haunting-ai-art-woman-explained.

Ryle, Gilbert. 2009. *The Concept of Mind*. New York: Routledge.

Sankar, Pamela. 2010. "Forensic DNA Phenotyping: Reinforcing Race in Law Enforcement." In *What's the Use of Race? Modern Governance and the Biology of Difference*, edited by Ian Whitmarsh and David S. Jones, 49–62. Cambridge, MA: MIT Press.

Schmitt, Philipp. 2020. *The Curse of Dimensionality* (online artwork). https://curseofdimensionality.science/ (accessed December 1, 2021).

Schmitt, Philipp, and Anina Rubin. 2020. "I Am Sitting in a High-Dimensional Room." Video documentation available at https://philippschmitt.com/work/i-am-sitting-in-a-high-dimensional-room (accessed July 5, 2021).

Schmitt, Philipp, and Stefan Weiss. 2018. "The Chair Project: A Case-Study for using Generative Machine Learning as Automatism." In *32nd Conference on Neural Information Processing Systems*, Montreal Canada. https://nips2018creativity.github.io/doc/the_chair_project.pdf.

Shaji, Appu. 2016. "Understanding Aesthetics with Deep Learning." *Nvidia Developer* (blog), February 29, 2016. https://devblogs.nvidia.com/understanding-aesthetics-deep-learning.

Silver, David, Aja Huang, Chris J. Maddison, Arthur Guez, Laurent Sifre, George Van Den Driessche, Julian Schrittwieser, Ioannis Antonoglou, Veda Panneershelvam, and Marc Lanctot. 2016. "Mastering the Game of Go with Deep Neural Networks and Tree Search." *Nature* 529 (7587): 484–89.

Simondon, Gilbert. 2009. "The Position of the Problem of Ontogenesis." Translated by Gregory Flanders. *Parrhesia* 7:4–16.

Simondon, Gilbert. 2015. "Culture and technics (1965)." *Radical Philosophy* 189: 17–23.

Simondon, Gilbert. 2017a. "The Genesis of Technicity." *e-flux*, May 2017. https://www.e-flux.com/journal/82/133160/the-genesis-of-technicity.

Simondon, Gilbert. 2017b. *On the Mode of Existence of Technical Objects*. Translated by C. Malaspina and J. Rogove. Minneapolis: Univocal.

Simondon, Gilbert. 2020. *Individuation in Light of Notions of Form and Information*. Vol. 2. Translated by Taylor Adkins. Minneapolis: University of Minnesota Press.

Simonite, Tom. 2018. "When It Comes to Gorillas, Google Photos Remains Blind." *Wired*, January 11, 2018. https://www.wired.com/story/when-it-comes-to-gorillas-google-photos-remains-blind/.

Simonite, Tom. 2021. "Now That Machines Can Learn, Can They Unlearn?" *Wired*, August 19, 2021. https://www.wired.com/story/machines-can-learn-can-they-unlearn.

Smilkov, Daniel, Fernanda Viégas, and Martin Wattenberg. 2017. "Visualizing High-Dimensional Space." Experiments with Google (interactive website), June 2017. https://experiments.withgoogle.com/visualizing-high-dimensional-space.

Srnicek, Nick. 2016. *Platform Capitalism*. Malden, MA: Polity Press.

Statt, Nick. 2020. "Google Expands AI Calling Service Duplex to Australia, Canada, and the UK." *The Verge*, April 8, 2020. https://www.theverge.com/2020/4/8/21214321/google-duplex-ai-automated-calling-australia-canada-uk-expansion.

Stein, Gertrude. 1998. "Plays." In *Writings, 1932–1946*, edited by Catherine R. Stimpson and Harriet Chessman, 244–69. New York: Library of America.

Stolcke, Andreas, and Elizabeth Shriberg. 1996. "Statistical Language Modeling for Speech Disfluencies." *IEEE ICASSP-96, 1996 IEEE International Conference on Acoustics, Speech, and Signal Processing* 1:405–8.

Stop Trump Coalition. 2020. Topple the Racists (website). https://www.toppletheracists.org/ (accessed January 26, 2021).

St. Pierre, Joshua Lane. 2018. "Fluency Machines: Semiocapitalism, Disability,

and Action." PhD diss., University of Alberta, Canada. https://era.library
.ualberta.ca/items/be106c3c-22f1-4ed8-9867-75c208552d42/view/da99776d
-2f67-41d4-ba30-cb7053f6ba3d/St%20Pierre_Joshua_L_201812_PhD.pdf.

Strickland, Eliza. 2023. "DALL-E 2's Failures Are the Most Interesting Thing About It." ieee Spectrum 22, December 22, 2023. https://spectrum.ieee.org/openao-dall-e-2.

Swanson, Steph Maj. 2022a. Loab, website, https://loab.ai/.

Swanson, Steph Maj. 2022b. Twitter, September 7, 2022. https://x.com/supercomposite/status/1567178300195323904?lang=en.

Swinford, Dean. 2001. "Defining Irrealism: Scientific Development and Allegorical Possibility." Journal of the Fantastic in the Arts 12 (45): 77–89.

Szegedy, Christian, Wei Liu, Yangqing Jia, Pierre Sermanet, Scott Reed, Dragomir Anguelov, Dumitru Erhan, Vincent Vanhoucke, and Andrew Rabinovich. 2015. "Going Deeper with Convolutions." 2015 IEEE Conference on Computer Vision and Pattern Recognition (CVPR). https://doi.org/10.1109/CVPR.2015.7298594.

Thompson, Jeff. 2018a. Interpolated Pebbles. Video. Uploaded to Vimeo November 18, 2018. https://vimeo.com/301470836 (accessed November 22, 2021).

Thompson, Jeff. 2018b. Pebble Dataset repository. https://github.com/jeff Thompson/Pebble-Dataset (accessed November 22, 2021).

Turing, Alan. 1950. "Computing Machinery and Intelligence." Mind 49:433–60.

Turk, Matthew, and Alex Pentland. 1991. "Eigenfaces for Recognition." Journal of Cognitive Neuroscience 3 (1): 71–86.

Ubbens, Jordan R., and Ian Stavness. 2017. "Deep Plant Phenomics: A Deep Learning Platform for Complex Plant Phenotyping Tasks." Frontiers in Plant Science 8 (July 2017). https://www.frontiersin.org/articles/10.3389/fpls.2017.01190/full.

Venn, Couze. 2006. "A Note on Assemblage." Theory, Culture and Society 23 (nos. 2–3): 107–8.

Verenikina, Irinia. 2010. "Twenty-First-Century Research." In Proceedings of ED-MEDIA 2010—World Conference on Educational Multimedia, Hypermedia and Telecommunications, edited by J. Herrington and C. Montgomerie, 16–25. Waynesville, NC: Association for the Advancement of Computing in Education.

Vinyals, Oriol, and Quoc Le. 2015. "A Neural Conversational Model." arXiv. Last revised July 22, 2015. https://arxiv.org/abs/1506.05869.

Voss, Peter. 2018. "No, Google Duplex Didn't Pass the Turing Test." Medium, May 11, 2018. https://uxdesign.cc/no-google-duplex-didnt-pass-the-turing-test-93f7235e6c40.

Waite, Thom. 2022. "Loab: The Horrifying Cryptid Haunting AI's Latent Space." Dazed, September 15, 2022. https://www.dazeddigital.com/life-culture/article/56964/1/loab-the-horrifying-cryptid-haunting-ai-latent-space-supercomposite.

Waters, Richard. 2023. "Man Beats Machine at *Go* in Human Victory over AI." *Financial Times*, February 19, 2023. https://arstechnica.com/information-technology/2023/02/man-beats-machine-at-go-in-human-victory-over-ai.

Weiss Keith M., Lambert Brian W. 2011. "When the time seems ripe: eugenics, the annals, and the subtle persistence of typological thinking." *Annals of Human Genetics* 75(3): 334–43.

Weng, Jingkai, Ding Yujiang, Chengbo Hu, Zhu Xue-Feng, Bin Liang, Yang Jing, and Jianchun Cheng. 2020. "Meta-neural-network for real-time and passive deep-learning-based object recognition." *Nature Communications* 11 (6309). https://www.nature.com/articles/s41467-020-19693-x.

Weizenbaum, Joseph. 1966. "ELIZA—A Computer Program for the Study of Natural Language Communication between Man and Machine." *Computational Linguistics, Communications of the ACM* 9(1): 36–45.

Wexler, James. 2018. "The What-If Tool: Code-Free Probing of Machine Learning Models." *Google AI Blog*, September 11, 2018. https://ai.googleblog.com/2018/09/the-what-if-tool-code-free-probing-of.html.

Whitehead, Alfred North. 1920. *The Concept of Nature*. Cambridge: Cambridge University Press.

Whitehead, Alfred North. 1978. *Process and Reality: An Essay in Cosmology*. New York: Free Press.

Wilson, Elizabeth A. 2010. *Affect and Artificial Intelligence*. Seattle: University of Washington Press.

Wong, Julia Carrie. 2019. "The Viral Selfie App ImageNet Roulette Seemed Fun—Until It Called Me a Racist Slur." *The Guardian*, September 18, 2019. https://www.theguardian.com/technology/2019/sep/17/imagenet-roulette-asian-racist-slur-selfie.

Wu, Shaohua, Xudong Zhao, Tong Yu, Rongguo Zhang, Chong Shen, Hongli Liu, Feng Li, et al. 2021. "Yuan 1.0: Large-Scale Pre-trained Language Model in Zero-Shot and Few-Shot Learning." arXiv. Last revised October 12, 2021. https://arxiv.org/abs/2110.04725.

Yang Kaiyu, Jacqueline Yau, Li Fei-Fei, Jia Deng, and Olga Russakovsky. 2021. "A Study of Face Obfuscation in ImageNet." arXiv. Last revised June 9, 2022. https://arxiv.org/pdf/2103.06191.pdf.

Yosinski, Jason, Jeff Clune, Anh Nguyen, Thomas Fuchs, and Hod Lipson. 2015. "Understanding Neural Networks through Deep Visualization." arXiv. https://arxiv.org/abs/1506.06579 (accessed February 17, 2019).

Zachnading. 2022. "AI-Generated Horror Cryptid Is Both Intriguing and Terrifying." *ebaum's World*, September 7, 2022. https://www.ebaumsworld.com/articles/ai-generator-creates-cryptid-and-shes-already-haunting-our-dreams/87264887.

Zeilinger, Martin. 2021. *Tactical Entanglements: AI Art, Agency, Intellectual Property*. Lüneberg: Meson Press.

Zeng, Wei, Xiaozhe Ren, Teng Su, Hui Wang,Yi Liao, Zhiwel Wang, Xin Jiang, et al. 2021. "PanGu-α: Large-Scale Autoregressive Pretrained Chinese Language Models with Auto-Parallel Computation." arXiv. Last revised April 26, 2021. https://arxiv.org/pdf/2104.12369v1.pdf.

Zylinska, Joanna. 2020. *AI Art: Machine Visions and Warped Dreams*. London: Open Humanities Press.

INDEX

Page locators in italics indicate figures, tables, and definitions

"accuracy," 48, 151–52
acoustics, 165–68
Activations (Bridle), 156
active shape models (ASMs), 108
activist and tactical media actions, 150, 154–55
"actual occasion," 74, 125, 139
actual technics, 37, 39, 62, 117, 177; as computational contingency, 24–27; and ML as *agencement,* 27–30, 77; and race, 80, 109
Aesthetica (Baumgarten), 7
aesthetics: "abstract" feeling, 73–75; aesthetic fact, 36, 73; commodifiable, 52; as domain of sensing relations, 8, 186n3; formal and sensorial approaches, 7–8; generalized equivalence accorded value, 71, 74–75; "glitch aesthetics," 65; "grain of computation," 155; heteropoietics (aesthetic diversity), 56–57, 69; and individuation, 9–10; moving difference, feeling of, 73–74, *74;* novelty, 69, 71–73. *See also* deepaesthetics
"Afro-now" AI, 38, 80, 109–13, 115

agencement, 28–30; and actual technics, 77; asignifying tendencies, machinic, 64–65; and category mistakes, 45–46; of computational conversations, 115; differential repetition of, 39, 157; errant, 38; as invisual, 79–80; and machinic heterogenesis, 48–49; *Not the Only One* as, 109–13; proto-*agencement,* 166; and quasi-qualitative relations, 89; and relationality, 29–30; and social effort of communities, 49–50; and statistical racism, 38, 84, 91, 95
"AI aesthetics" (Manovich), 56–57
AI Fairness 360 tool kit (Linux), 85
Alcine, Jacky, 85–88, *86, 87*
Alexa, 114–15
AlexNet, 41, *55,* 65
algorithms, 18; "Black technical object," 93–94; compression, 166–67; data relations put into inverse relation with, 163–64; data structures distinguished from, 83; evolutionary, 65–67; fairness, 85; nearest-neighbor, 163;

algorithms (*continued*)
    and occurrent learning, 76; as a priori axiom, 24–25; in process of becoming, 16; racial bias in, 85–88, *86, 87*; t-SNE (t-distributed stochastic neighbor embedding), 167–68; "unreason" of, 154; and Western conceptions of seeing, 80. *See also* input data (noise)
"Alice" and "Bob" bots, 63–64, *64*, 75, 118
allagmatics, 39, 157–59, 166, 169, 171, 173, 178. *See also* machine unlearning (MUL)
allele frequencies, 103
Alli, Kabir, 86, 193n5
allopoiesis, 37, 51
Alpaydin, Ethem, 62
Alpern, Emma, 117, 118
AlphaGo, 18, 59, 190n14
Amaro, Ramon, 89, 93, 95, 108
Amazon, 170; conversational AI, 114–15; Mechanical Turk, 149
Amoore, Louise, 154
analogical operativity, 156–58; in irrealist artworks, 159–65
Anderson, Edgar, 81, 159–60
Andrejevic, Mark, 129
architecting, 93
Artbreeder, 71, 192n25
artful techniques, 4, 20, 36–39, 80, 180, 187n7; analogical operativity, 156–58; art and "not-art," 72, 191n21; high dimensionality in, 159, 166–69; imitation game, 122–23; irrealism, 159–65, *162*; and latent spaces, *46*; modulating of conversational AI, 142–46. *See also* machine unlearning (MUL)
artificial general intelligence (AGI), *134*, 134–38, 196n9; defined, *135*; unable to understand performance, 134–35, *135*
artificial intelligence (AI): AI-cryptids, 1–4, *2*; allopoietic dimension of, 37, 51; binary operations, 17, 85, 178; Blackness and/as a multiplicity of, 109–13; as differencing machine, 51; environmental problems both produced and solved by, 170; ethical pitfalls of, *45*, 150–51, *153*, 171; exploitative labor practices, 150–51; as human-machine ensemble, 23; military uses of, 128, 131, 195–96n6,

196n7; as at odds with itself, 36, 51, 110, 178; operational strategies, 155; "race problem" of, 86–87, 89, 111; risk-averse cultural imaginary for, 49; structure versus operation, 156–59; technogenesis of, 132. *See also* category mistakes; computational experience; creativity of AI; human-machine relations; machine learning (ML)
art practices, 36, 155, 187n7, 191n21
arXiv, 40–41, 188n2
asignifying tendencies, 64–65, 117, 142, 167; machinic, 64–65; neurodiverse, 117; and prediction of signifying content, 130–31
Asunder (Brain), 170–73
autonomous systems, 17, 30, 141
autopoietic vectors, 50–52, 85, 88

back propagation, 17, 18, 93
Banerjee, Imon, 104
Baumgarten, Alexander, 7
Bayesian clustering techniques, 102–3
Belhumeur, Peter N., 107
benchmarks, 41, 49, 65, 69, 188n9; and artful techniques, 159–60, 163, 165; human capacity as, 51, 68, *134*, 135; Turing test, 120. *See also* datasets; Iris dataset
Bengio, Yoshua, 10–11, 71
Benjamin, Ruha, 92–93, 98
Bergson, Henri, 192n1
Berkeley Restaurant Project, *128, 129*
Berry, David, 155
Bethge, Matthais, 53, 58, 190n19
big data, 25, 60, 127. *See also* graphics processing unit (GPU)
BINA48, 109–11, 194n15
biometric recognition, 6, 99
biopolitics, 91–92, 193n6
Bitcoin, 164
black box, 3, 16, 23, 155–56
blackness: "Afro-now" AI, 38, 80, 109–13, 115; and algorithmic racial bias, 85–88, *86, 87*; Black and People of Color communities, 38, 89–90, *90*, 109; and multiplicity of artificial intelligences, 109–13. *See also* race; racialization; racism
Blais, Tim, 166
Bodmer, Walter, 100

Boney, Robbie, 43
boyd, danah, 25
Brain, Tega, 170–74
Brandom, Russell, 125
Bridle, James, 156
Bucher, Taina, 16, 187n6
Buolamwini, Joy, 84

Cain, Joe, 101
Cambridge Language Research Unit, *145*
capitalism, 30, 155, 165; platform, 119, 132, 160, 170, 185–86n2
category mistakes, 40–44; "accurate" error, 67; bot training experiment, 63–65, *64*; of computational vision, 38–39; diagnostic potential of, 47–48; duck-rabbit illusion problem, 41, *42*, 43, *44*, 47, 72, 75, 188–89n3; and facial recognition, 148–49; by "foreigners," 42–43, 47–48; and high-dimensionality, 168; and machine unlearning, 165, 180; and meta-categorical concepts, 42–43; move to computational heterogenesis from, 48–51; and optical illusions, 41, *42*, 43; "university," image problem, 42, 43–44, *44*. *See also* computer vision; image recognition and generation; neurodiversity
causality: mechanical, 91; recurrent, 32, 34–3513, 187n9
central processing unit (CPU), 59, *60*
Chambers (Lucier), 166
The Chair Project (Schmitt and Weiss), 18–20, *19*
chatbots: "Alice" and "Bob," 63–64, *64*, 75, 117; ChatGPT, *121*, 127, 179; ELIZA, *120*, *121*, 122, 195n4. *See also* conversational AI; Not the Only One (N'TOO, Dinkins)
childhood development, 138, 196n10
Chun, Wendy, 92
classificatory schemas: color line produced via forms of, 79; labels and classes, 86–88, *87*; racism made operational by, 86–87, *87,* 193n5; typological, 89, 91–93, 108
Clayton, Aubrey, 91
co-composition, 18–20, *19*, 36, 77, 83, 113, 146
cognitivist/constructivist theory of mind, 136–38

color: color line, 79–80, 100; in data visualization, 78–79; and facialization, 99; hue, 79; indeterminacy of, 93; not perceptible to ML, 88–89; worlding of, 88–89. *See also* racialization
Combes, Muriel, 168
comma-separated values (CSVs), 79
"common sense," 136–38
Community Earth System Model, 170–71
compression algorithms, 166–67
computation: axiomatic nature of, 24; as differencing, 17; digital, 25, 76, 155; at nonhuman scales, durations, and dimensions, 9, 11–12; as phenomenotechnics, 8; statistical, 3, 15, 25, 27, 38, 105
computational experience, 5–10; and dimensionality reduction, *14*, 15–16
"computational idealism," 24
computer vision, 26, 156; category mistakes of, 38–39; constraining, 52–58; and data accuracy, 151–52; as difference from human vision, 67; "fooling," 63–68, 75; as "hallucination," 9, 50–51, 163; human capacities replicated by, 68; ImageNet dataset project, 49; pattern recognition research, 65. *See also* category mistakes; image recognition and generation
ConceptNet, 142, 197n16
The Concept of Mind (Ryle), 42
concepts, and human experience, 136–37
concrescence, 174
conjunction, 5, 7, 20
Constantino, Christopher, 133
constructivism, 136–37
context-centric neural architectures, 126–27, 131–32
continuity and discontinuity, processes of, 5, 20, 72–73, 75, 138–39; and deepaesthetics, 10–11, 27, 30; in early layers, 76–77; and race, 105–9
control, societies of, 177
conversation, generalized domain of, 139–42, 144
conversational AI: actual technics of, 117; artful techniques for modulating, 142–46; context-centric neural architectures, 126–27, 131–32; creolization, 127, 134; fluency-disfluency relation, 133–34, 142;

conversational AI (*continued*)
and general intelligence, 118–19; materialities of, 131–34; narrowness and generalization, 125–26, 135, 194n4; from "natural" conversation to general AI, *134*, 134–38; and neurodiversity, 118, 133; and Pask's cybernetics, 139–42, 196–97n12; playing at the conversation game, 119–25; question-and-answer structure, 122, 127; statistical "natural" conversation, 125–31; task-oriented dialogues, limitation of, 114–15, 118–19, 125–26, 136; technical ensembles required by, 115–16; tone, importance of, 120, *121*, 122; turn toward "naturalistic" language flow, 117; wayward tendencies, 115, 123, 144. *See also* chatbots; natural language processing (NLP); neurodiversity; relationality

Conversations with Bina48 (Dinkins), 110–11

Conversation Theory (Monin), 119, 142–46

convolutional neural networks (CNNs), 41, 97; defined, *55*; style transfer, 53–58, *54*, 63, 69, 71, 75, 189n12, 190–91n19; and vectorization of race, 98–99

correlation, 90–91, 103

Courville, Aaron, 10–11, 71

Cox, Geoff, 80

Crawford, Kate, 25, 147–50, 196n8, 198n2

creative adversarial network (CAN), 68–74, *74*, 192n26, 192n27; art historical taxonomy, 69

creativity of AI, 7, 21, 23, 52; "computational creativity," 67, 68–77; and ILSVRC project, 9, 50–51

creolization, 43, 127, 134

critical data archaeology, 150

Critique of Judgment (Kant), 7

cryptids, 1–4, *2*

cultures: of computation, 84; of platforms, 52, 86–87, 89; of sensing, 8; visual, 53–54, 57, 86, 165, 191n19

Curse of Dimensionality (Schmitt), 168–69, 199n10

cybernetic recurrent causality, *31*, 31–32

cybernetics, 36, 174; negative feedback, *31*, 32; Pask's, 139–42, 144–45, 196–97n12

DALL-E, 3, 43, 189n5

data: discreteness, myth of, 152–53; labels and classes, 86, *87*; relational opticality of, 81–84; as value, 159, 163

data accuracy, 151–52

data model, 149

data objects, 83, 106

data ontology, *149*

data provenance, 131–32, 150, 154

datasets: accuracy of, 48, 151–52; algorithms put into inverse relation with, 163–64; automated scraping for, 151, *153*, 160, 191n22; *Iris* data, 81–84, *82*, 160, 198n7; manually crafted, 157, 159–65, *162*. *See also* benchmarks

data structure, 81, *83*

data table, 15, 80–84, *82*, *83*, *95*, 160, 198n7

data visualization, 78–79

deepaesthetics, 67, 179; alternate, 18, 37–38, 51, 71–72, 80, 158, 180; banal, 48; continuities and discontinuities, 10–11, 27, 30; experiencing, 4–10; and race, 80, 85, 109. *See also* aesthetics

DeepFace, 94, 97

deepfakes, 2–3, 11, 191n24; "real fakes," 71

deep learning, 3, 7, 185n2; as avolumetric, *12*, 12–13, 28, 169; and conversational AI, 125–26; "deep," defined, 10, 11–12; depth of, 10–16; general intelligence, 118–19; quantitative problem of, 12–13; representationalist paradigm, 9; surface-generated depth, 15. *See also* machine learning (ML)

DeepMind, 59, 71

deep neural networks (DNNs), 11–12, 15–16, 65, *66*, 134, 4167

DeepQ&A, 113

Deep Swamp (Brain), 173–74

Defense Advanced Research Projects Agency (DARPA), 128

DeLanda, Manuel, 29

Deleuze, Gilles, 28–30, 99, 166, 197n13; modulation, 177; "outside of language," 117; "signaletic material," 197n14

Desrosières, Alain, 90–91

differencing, 17, 21, 23, 51; art and "not-art," 72, 191n21; and conversational AI, 126; moving flows of images, 73–75, *74*

diffusion models, 46–47
digital technology, 155
dimensionality: "curse" of, 168–69; increase in, 166–67; n-dimensionality, 13, *14*. *See also* high dimensionality
dimensionality reduction, 15–16, 18, 29, 37, 56, 93; defined, *14*. *See also* linear discriminant analysis (LDA); principal component analysis (PCA)
Dinkins, Stephanie, 38, 80, *90*, 109–13, 115, 117, 194n15
discreteness, 75–77
discriminant analysis, 52–53, 65–67; racist legacy of, 93. *See also* linear discriminant analysis (LDA)
disequilibrium, 118
disfluency, 38–39, 120; as condition for fluency, 117, 132, 133, 142; opening onto, 127–28; pathologized for human speakers, 116. *See also* language; neurodiversity; stuttering
diversity, aesthetic, 56–57, 69
domain specificity, 135, 136
downsampling, *46*, 54, 62, 115, 190n12
drifts (*dérives*), 115, 143–44
duck-rabbit illusion problem, 41, *42*, 43, *44*, *47*, *72*, *75*, 188–89n4

Ecker, Alexander S., 53, 58, 190n19
edges (links), 5, 11, 88, *145*, 156
eigenfaces, *96*, 97–98, 190n18
eigenvector analysis, 97
elaboration, 64, 139, 143–44
Elgammal, Ahmed, 68–69, 72, 74, *74*, 192n26
ELIZA, *120*, *121*, 122, 195n4
environmental problems, 8, 170–74
eugenics, 87; as biopolitical, 91–92, 193n6; correlation theory, 90–91; and linear discriminant analysis, 80, 105–9; "practical," 99–100; and principal component analysis, 80, 99–100; race and population, understanding of, 100–103; and racist imaginary of platform cultures, 89; as social machine, 53; statistics-eugenics nexus, 38, 80, 91–93, 95, 102–4; stock as unit of concern, 101. *See also* racialization
Everalbum, *153*

evolutionary algorithms, 65–67
experience: aesthetic "feeling" as, 73; change changing, 75, 119; computational, 4–10; imperceptible statistical relations in, 5–6; intense, 36, 73; Jamesian, 5, 8–9; lived, 5, 86–87; machine learning, 6, 18; as measured phenomenon, 17; neurodivergent, 118; novel, via ML, 69, 71–73; as open relationality, 20–21, 23–24. *See also* processuality
Eye/Machine (Farocki), 195–96n6

Facebook: bots experiment, 63–65, *64*; DeepFace, 94, 97; Fairness Flow, 85
facial image generation, 69–71; celebrities dataset, 71, 72
facial recognition, 7, 84, 190n18; by active shape models (ASMs), 108; datasets deleted, 150; eigenfaces, *96*, 97–98, 190n18; and machine unlearning, *153*; "person" classes, 147–49, *149*, 151; and racial profiling concerns, 98
Fairness Flow (Facebook), 85
Farocki, Harun, 195–96n6
Fazi, Beatrice, 7–8, 17, 24–25, 75–76
features, *13*, 21; equilibrium of, 70; feature maps, 53–57; feature space, 57–58; image style, 53–54; "principal," 95
feedback loops, 34
Fisher, Ronald A., 52–53, 90, 91, 193n7; eugenics work of, 100–104; *Iris* dataset, 81–84, *82*, 94, 160, 198n7; and linear discriminant analysis, 105–9
Flickr, 174; Flickr-Faces-HQ (FFHQ), 69–70, 191n22
"fooling": in computer vision, 63–68, 75; in conversational AI, 120, 122
forgetting, 151, 198n4
formalism, 7–8, 52, 149
Foucault, Michel, 48, 193n6
Fuller, Matthew, 8–9, 186n3

Galton, Francis, 91, 101
Gatys, Leon A., 53, 55, 58, 190n19
Gebru, Timnit, 84
generalizability, 67, 71, 93, 137; and Pask's conception of conversation, 139–42, 144–45

INDEX 225

generalized intelligence, 118–19, 138, 196n9
generative adversarial network (GAN), 18–20, *19*, 41, 59, 179, 191n23; and climate modeling, 172; creative adversarial network (CAN), 68–74, *74*, 192n26, 192n27; defined, *70*; generator and discriminator networks, 69–73, *70*, 172; and manually crafted datasets, 159–61; relative instability of, 164
generative image models, 4, 45, 55; ILSVRC, 9, 49–50, 150, 189n7
genetics, 37, 101; STRUCTURE analysis software, 102–4, 193n11
Glushko, Robert, 69
Gödel, Kurt, 75
Goodfellow, Ian, 10–11, 71
Goodman, Andrew, 172–73
Google: AI blog, 126; Assistant, 119–21, 132; Brain, 41, 68; DeepDream, 49–51, 71; Duplex, 119–22, 125–27, 131–33, 135, 141; Inception, 9, 50; I/O event, 119–21; Photos, 85–88, *86, 87*; What-If developer toolkit, 85
GPT (Generalized Processing Training), 11, 177–81, 195n5, 196n11
graphics processing unit (GPU), 59, *60*, 77, 186n2, 188n9; defined, *61*
Grundtmann, Naja, 58, 189–90n12
Guattari, Félix, 28–29, 30, 64, 99, 166, 187n8; machinic heterogenesis, 48–49, 51; technicity of elements, 37

Haldane, J. B. S., 101
"hallucinations," 5, 50, 68, 71, 163
Hansen, Mark, 8
Harney, Stefano, 110
Hartman, Saidiya, 79
Harvey, Adam, 150
Herndon, Holly, 35–36, 113, 188n10
Hertzmann, Aaron, 71
Hespanha, Joao P., 107
heterogenesis, machinic, 48–49, 51, 71
heteropoietics (aesthetic diversity), 56–57, 69
high dimensionality, 13, *14*, 22, 47, *61*; in artful techniques, 157, 166–69; and LDA, 107–8; and principal component analysis, *96, 97*. *See also* dimensionality

Hui, Yuk, 187n9
human intelligence taskers (HITs), 18, 149–50
"Human Language Technology" (DARPA), 128–31, *130*
human-machine relations, 23, 32–35, 51, 87; *The Chair Project* (Schmitt and Weiss), 18–20, *19*; conversational, 38–39; input-correlation-output cycle, 32; schematization in, 34–35

I AM AI (Southern), 34–35
I Am Sitting in a High-Dimensional Room (Schmitt and Rubin), 165–68
I Am Sitting in a Room (Lucier), 165–67
I Am Streaming in a Room (Blais), 166
IBM, 85, 127, 128
Ikoniadou, Eleni, 167
image generators, DIY, 50
image grids, 156
image kernels, 54, *55*
ImageNet, 41, 55, 65, 147–50, *149*, 160, 197n1; dataset project, 49; labels and classes in, *87*
ImageNet Large Scale Visual Recognition Challenge (ILSVRC), 9, 49–50, 150, 189n7
ImageNet Roulette, 147–50, *148*, 151, 198n2
image recognition and generation, 23, 26, 37; developments in, 40–41; eugenics genealogy of, 91; large-scale image models, 98. *See also* category mistakes; computer vision
image style, 53–54. *See also* style transfer
"imitation game," 122
immanent critique, 157–58
incalculability, 172–73
incomputability, 16–17, 75, 77
indeterminacy, 16–17, 26–27, 30; of color, 93; and environmental engineering, 174; margin of, 35–37, 51, 62–63, 141, 158; and music improvisation, 35; quantitative, as ground for novelty, 75
individuation: aesthetic, 9–10; of "Black technical object," 93–94; and conversation, 134, 140–42; of living beings, 174
industrialized AI, 170–75; environmental solutionism of, 170–71

input-correlation-output cycle, 32
input data (noise), 4, 9–13, 47, 193n9; "prior constraint," statistical, 9–10, *13*, 50
intelligence, generalized, 118–19; artificial general intelligence (AGI), *134, 134*–38, *135*, 196n9
Interpolated Pebbles (Thompson), 163–65
invisuality, 79–80, 171–72. *See also* visualizations
Iris dataset (Fisher, Anderson), 81–84, *82*, 90, 94, 159–60, 192n3, 198n7
irrealism, 159–65

James, William, 5, 8–9, 73, 104–5, 154, 188n3, 193n12; "common sense," view of, 137–38; intimacy, concept of, 20; openness of experience, 20–21, 23–24; process thinking, 5, 20–21, 23; radical openness, 21, 137
Jelinek, Frederick, 127, 128
Jolicoeur-Martineau, Alexia, 68
Joppa, Lucas, 171

Kahn, Douglas, 166
Kant, Immanuel, 7
knowing, automation of, 25
knowledge production: politics of, 150–51; as structure of AI, 156
Knuth, Donald, 187n7
Kriegman, David J., 107

labels and classes, 86–88, 105; definition, *87*; and LDA, 107; "person" classes, 147–49, *149*, 151; and predictions in policing and profiling, 92–93
Lambert, Brian, 102–3
language: asignifying potentialities, 64–65, 167; exteriority of normalized ("fluent") speech, 39, 117–18; machine communicability, 64, *64*; "natural ambiguity" of, 127; neurodiversity of, 38–39, 116–17, 132–33, 142; phonic fluency, 124–25; speech pathology, 38, 132–33; stuttering, 39, 116–17; subject of communication, 129, 133–34. *See also* category mistakes; conversational AI; disfluency; large language models (LLMs); natural language processing (NLP)

language models (LMs), 127, *129*
Laplace, Jules, 150
Lapoujade, David, 20–21
large language models (LLMs), 3, 11, 98, 135, 196n11; and speech recognition, 128, 130. *See also* text-to-image models. *See also* language
latent space, 4, 19, 46–47, 60–63, 190n17; artful experimentation in, 19, 46, 163–64, 168; defined, *46*; nearest-neighbor algorithms, 163; and race, 100, 190n18. *See also* generative adversarial network (GAN); space-time
layers, 10–11; defined, *13*; early, and processual continuity, 76–77; epistemic work of, 27–28; "on top of" metaphor, 12, 55; pooling, 53, 55, 58, 105, 152; visual representations of, *12*. *See also* neural network architectures
Lesson One (Prévieux), 123–25, 144
Leviathan, Yaniv, 126, 132
Liddell, Patrick, 166
linear discriminant analysis (LDA), 37, 52–53, 193–94n13; defined, *106*; eugenics linked with, 80, 105–9; as qualitative, 107–8; and statistical racism, 80, 94–98
lived experience, 5, 86–87
Loab (cryptid), 1–4, *2*, 21
Lucier, Alvin, 165–67
Lucier, Mary, 166

Macauley, William, 166
machine learning (ML), 2–4; actual technics as computational contingency, 24–27; *agencement* of, 28–30; as "biased," 26; color not perceptible to, 88–89; eugenics genealogy of, 79, 90–91; irrealism of, 159–65; James's philosophy of experience applied to, 5, 8–9; machinic universe of, 27–30; management techniques, 13, 26–27, 100, 180, 196n9; as occurrent learning, 17, 76; potential inventiveness of dismissed, 48; quantity and quality in, 6–7, *14,* 15; radically empirical experience for and of, 16–24, 104–5; recursive and modulating functioning of, 16–17; sensibilities specific to, 18; Simondian technics of, 30–37;

machine learning (ML) (*continued*)
  speculative, 165–69; as statistical computation, 25; updating of, 16, 76, 187n6; vectorization, 15, 18, 22, 46, 60–62, 97, 190n18. *See also* artificial intelligence (AI); deep learning; errant mode of ML; human-machine relations; neural network architectures
machine unlearning (MUL), 151, *152*; artworks and AI sensorium, 154–55; defined, *153*; speculative operativity, 170–75
machinic perception, 57–58
Mackenzie, Adrian, 15, 25–26, 58, 103, 107, 188n1, 188n3, 191n20, 192n1
Manning, Erin, 36, 88–89, 94, 133, 134, 186n5, 187n7
Manovich, Lev, 56–57
Markov chain Monte Carlo (MCMC), 102–3
Massumi, Brian, 119, 123, 157, 158, 186n5, 189n4
mathesis, statistical, 25, 48, 51–53, 76, 79, 105
Matias, Yossi, 126, 132
Mbembe, Achille, 112
McCarthy, John, 136
McTear, Michael, 115
medical analysis, 84, 104, 193–94n13
metastability, 118, 141, 174, 194–95n2
Microsoft "Planetary Computer," 170–71
Microsoft Research, 85
micro-spatiotemporalities, 59, 60, 77
Midjourney, 3, 23, 43, *45*
military uses of AI, 128, 131, 195–96n6, 196n7
Mitchell, Tom, 6, 76
MNIST handwriting dataset, 41
modulation, 142–46, 177–81
Monin, Monica, 119, 142–46, 197n15
more-than-human registers, 20, 37, 51, 171–72, 186n5
Morrison, Toni, 111
Mosaic Virus (Ridler), 159, 161, 164–65
Moten, Fred, 90, 110
motifs, 15, 53–54, 57, 74, 152
moving difference, aesthetic feeling of, 73–75, *74*
Murphie, Andrew, 73
music, 30–35; AIVA music composition tool, *33*; Auto-Tune, 30–32, *31*, 34
Myriad (Tulips) (Ridler), 159–61, 164–65

Naitzat, 15
natural language processing (NLP), 114, 116; Berkeley Restaurant Project, *128, 129*; context-centric neural architectures, 126–27, 131–32; defined, *121*; FPs (filled pauses), 130–31; imitative functionality, 122; increasing quantities of training data and parameters, 138; "natural" language, 38–39; n-grams (*n-1* words), 127, *129*; and recurrent neural networks (RNNs), *33*, 131; semantic networks, 143, *145*, 197n16; statistical turn in, 127–28; SWITCHBOARD dataset, 128–32, *131*, 142, 196n6. *See also* conversational AI
negative feedback, *31*, 31–32
neural network architectures, 7, 174–75; black box, 3, 16, 22, 155–56; deep learning, 10–16; nodes, *22*; nonlinear functioning of, 16–18; recurrent neural networks (RNNs), *33*, 34, 113, 131; representationalist paradigm of (visual) perception, 9; subspaces, 99; visualizations of, *14*, 62, 78–79. *See also* layers; machine learning (ML)
neurodiversity, 194n1; AI's potential for, 123, 133–34; of language, 38–39, 116–17, 132–33, 142. *See also* category mistakes; conversational AI; stuttering
n-grams (*n-1* words), 127, *129*
Nguyen, Anh, 65–68, *66, 67*
nodes, *22*
nonlinear recursions, 16–18
"Nonspeaking, 'Low-Functioning,'" (Sequenzia), 194n1
"normal distribution curve," 100
Not the Only One (N'TOO, Dinkins), 89, 109–13, 115, 117
"novel consequent," 73
novelty, 71–77; and generalization, 138; as variability within narrow distribution of the same, 71–72
Nvidia, 52, *70, 71, 72*

occurrent learning, 6–7, 17, 76
Oliver, Julian, 170
OpenAI, *45*, 179, 186n2, 189n4, 196n11
openness, 21, 30, 35, 67, 137, 141
operational images, 129, 195–96n6

operationalization, 61, 83; of race, 38–39, 80, 83, 90–94, 98
Operational Land Imager, 171
operative tracing, 156–58, 160
operativity, 6; analogical, 156–58; analogical, and irrealist artworks, 159–65; speculative, 170–75
optical illusions, 41, *42*, 43
Ousley, Stephen D., 105
Overstory, 170, 199n12

Paglen, Trevor, 147–50, *148*, 196n8, 198n2
Panofsky, Erwin, 69
parallel processing, 59–60, *61*, 188, 196n9
parametrization, 76, 104, 105, 108, 151
pareidolic images, 50–51, 68
Parisi, Luciana, 16–17, 67
Pask, Gordon, 139–42, 144–45, 196–97n12
Pasquinelli, Matteo, 99
Pearson, Karl, 90–91, 99–100, 108
Pebble Dataset (Thompson), 161–63, *162*, 165
Peirce, Charles Sanders, 62
Pelrine, Kellin, 190n14
Pentland, Alex, 97, 98
perception: computational experience as beyond, 8–11, 19; machinic, 57–58; visual, 9, 43, 50–51, 67, 80, 189n4
"person" classes, 147–49, *149*, 151
photorealism, 63, 71–72
Photoshop (Adobe), 56
phylum, machinic, 37, 41, 111, 174, 187n9
Picbreeder, 71, 192n25
Pichai, Sundar, 120
"Planetary Computer" (Microsoft), 170–71
plasticity, 142, 154, 167
platform capitalism, 119, 132, 160, 170, 185–86n2
platform-enabled ML, 38, 60
platforms: cultures of, 52, 85–87, 89; racism made operational by, 86–87, *87*
playfulness, 122–24; deploying, 157
Polaroid Slides (Lucier), 166
population, 25, 100–103
Pop-Up definitions, 186n5; artificial general intelligence (AGI), *135*; convolutional neural networks (CNNs), *55*; data ontology, *149*; data structure, *83*; dimensionality reduction, *14*; generative adversarial network (GAN), *70*; graphics processing units (GPUs), *61*; labels and classes, *87*; latent space, *46*; layers in neural networks, *13*; linear discriminant analysis (LDA),*106*; machine unlearning (MUL), *153*; natural language processing (NLP), *121*; negative feedback or cybernetic recurrent causality, *31*; n-grams, *129*; principal component analysis (PCA), *96*; recurrent neural networks (RNNs), *33*; semantic networks, *145*; text-to-image model, *45*; weights and biases, *22*
prediction, 192n26; in Black policing and criminal DNA profiling, 92–93; and capitalist agenda, 30; classical statistical methods replaced by, 105–6, 108; and contingency, 68–69; genericism in, 92–93; and layers, 11; and odd spacetime of ML's autopoiesis, 58–59; social, cultural, and political arrangements of, 23; sociotechnics of, 26–27; and vectoralization, 15
Prévieux, Julien, 123–25, 144
Princeton University, 144, *145*, 150
principal component analysis (PCA), 18, 37; defined, 96; eigenfaces, *96*, 97–98; eugenics linked with, 80, 99–100; Pearson's proposal of, 94, 95, 108; "sparse," 193n9; and statistical racism, 37, 94–98. *See also* dimensionality reduction
prior constraint, statistical, 9–10, *13*, 50
Prisma app, 190–91n19
probabilistic logic, 25–26, 45–46, 48, 76, 102; and conversational AI, 126–28
probability, 58; automation of, 25–26; diffusion models, 46–47; shaping of by LDA, 107; word prediction, *121*, 127–28, *128*, 129
process philosophy, 8, 16, 80, 137
process probes, 4
process thinking, 5, 20–21, 23
processuality, 178, 181; aesthetic feeling specific to, 75; bifurcated, 153–54; of modulation, 142; and race, 93; technical ensembles opened to, 48–49. *See also* experience
Project al-Khwarizmi (Dinkins), 109–10

INDEX   229

prompts: across model and designer, 18–20, *19*; negative prompting, 3–4, 158

qualitative processes, 13–16, 18, 23, 51, 128, 159, 165; and "abstract" image, 73–75, 77; and continuity, 20, 73; and "feeling," 73; and latent space, 60, 63; and LDA, 107–8; and racialization, 89, 92. *See also* vectorization

quantitative processes, 7, 12, *14*, 15–18, 23, 24, 75–77, 161, 165; and natural language processing, 118, 127, 130, 136; and race, 78, 92–94, 97, 101–2, 105, 107–8

race: after computation, 93; of biomedical images, 104; data, whiteness, and Blackness, 85–90; eugenic project of constraining variability, 91, 93, 100, 108; operationalization of, 38, 80, 84, 92, 98; operationalization of, through the statistical, 39, 90–94; as plastic and discrete, continuous and discontinuous, 105–9; as radical empirical experience, 104–5; and sociotechnicalities of machine learning, 38, 85, 95, 100, 110–11, 115; and statistics-eugenics nexus, 38, 80, 91–93, 95, 102–4; and stock as term, 101; as technology or technique, 92–94; vectorization of, 91–92, 98–105, 190n18

racialization, 177–78; by active shape models (ASMs), 108; and becoming-operational of ML, 85; and category mistakes, 37–38; fairness algorithms, 85; and genetics, 102–3; statistical logic of, 26, 37, 79–80. *See also* eugenics

racism: actual technics of, 37–38; and *agencement*, 38, 84, 91, 95; AI's "architecture" of and white sociality, 92–93; biological, 193n6; and classificatory regimes, 79, 86–87, *87*, 193n5; debiasing solutions, 85, 87–88, 105, 109, 150, 154; figuring, 94–98; historical continuity of, 79, 88; made operational by platforms, 86–87, *87*; "race problem" of AI, 86–87, 89, 111; racial profiling, 98

Ramesh, Aditya, 43

"real fakes," 71

recurrent causality, 32, 34–35, 113, 187n9

recurrent neural networks (RNNs), *33, 34*, 113, 131

recursivity, 187n9; nonlinear, 16–18

relationality: and *agencement*, 29–30; computational, 18; of computational conversations, 115; of data, 13; and data accuracy, 151–52; of experience, 9; fluency-disfluency, 133–34; latent space of, 168; and machine unlearning, 151–52; of node, 22; nonrelation of, 23; open, 20; opticality of data, 81–84; performance of by ML, 76–77; quasi-qualitative relations, 89; of race, 104–5; "relational network," 139; of resonance, 166; sociability, 139–40; statistics as topology of, 81. *See also* conversational AI

rememory, 111–12

repetition, 39, 74–75, 125, 197n13

representationalism, 9–10, 50, 136–37, 159, 168, 180

resonance, 166–68

"rhythmic event," 167

Richens, Richard, *145*

Ridler, Anna, 159–65, 198n6

Rombach, Robin, 46

Rothblatt, Bina Aspen, 194n15

Rubin, Anina, 165–68

Ryle, Gilbert, 41–43, 47–48

sampling, 25–26, 36, 43, 129–30, 191n21; and race, 86, 88, 103

Sankar, Pamela, 98

satellite images (Landsat images), 171, 199n13

schiz, 134, 138

Schmitt, Philipp, 18–20, *19*, 165–68, 199n10

semantic networks, 143, *145*, 197n16

sense making, 8, 146

sensing, 186n3; computational, 8–9, 19, 170–72; cultures of, 8; human, 10–11

sensorium, AI, 154–55

Sequenzia, Amy, 194n1

SERI MATS (Stanford University), 195n5

Simondon, Gilbert, 30, 32, 94, 166, 174–75, 187–88n9; allagmatics, 39, 157; individuation, view of, 140–41; margin of indeterminacy, 35–37, 51, 62–63, 141, 158;

metastability, 118, 141, 174, 194–95n2; operative tracing, 156–58
Sjölén, Bengt, 170
sociability, 139–40
social media, 61, 68, 71, 147
sociotechnical ensembles, 29, 38, 51, 110; and artful techniques, 154, 156, 158, 165
sociotechnicalities of machine learning, 36–39, 57–59, 77, 132; and actual technics, 24; for automated predictability, 30, 158, 173; and category mistakes, 40, 48–49, 51–52; classificatory schemas, 69; constraints on ML, 52–58; and geoengineering, 173; and operativity of ML, 156–57; of predictive computing, 26–27; and race, 38, 85, 95, 100, 110–11, 115; topologies of surface-generated depth, 15
Southern, Taryn, 34–35
space: feature space, 57–58; nonphysical, 166–67; three-dimensional, 165–68; topological, 12, 13, 15. *See also* latent space
space-time, 5; micro-spatiotemporalities, 59, 60, 77; oddness of ML's, 58–63. *See also* latent space
Spawn, 35–36, 113, 188n10
speculative machine learning, 165–69
speech pathology, 38, 132–33, 194n1
speech recognition, 127–28
Stable Diffusion, 3, 43, *44, 72*
standardization, 32–33, 49, 163, 188n9
The Starry Night (Van Gogh), 56, 58
statistics: black-and-white enumeration of, 79, *82*, 90; data visualization, 78–79; logic of racialization, 26; multivariate classification, 105; and natural language processing, *121*; prediction as replacement for classical methods, 105–6, 108; probabilistic logic transposed from, 25–26; race operationalized through, 39, 90–94; and racist *agencement*, 38, 84, 91, 95; statistics-eugenics nexus, 38, 80, 91–93, 95, 102–4; as topology of relational surfaces, 81. *See also* racism, statistical
Stein, Gertrude, 117
St. Pierre, Joshua, 133
streaming, 166–67
structuration, 39, 84

STRUCTURES analysis software, 102–4, 193n11
stuttering, 39, 133, 178, 180; as disfluency, 116; as neurodivergent speech, 116–17; poetic, 117–18. *See also* neurodiversity
StyleGAN, *70,* 94
style transfer, 63, 71, 180, 190–91n19; aesthetics of, 75; emulating and deviating from, 69; feature maps, 53–57; style-content relation, *54,* 58, 189–90n12; "style" identified, 53–54; Van Gogh Tübingen canal scene, 56, 58
stylistic ambiguity, 69
subspaces, 99
The Sun, 132
surface, 8–10, 15, 81; hypersurface, 28. *See also* layers
Swanson, Steph Maj, 1–4, 185n1
SWITCHBOARD dataset, 128–32, *131,* 142, 196n6
synthesis, 10, *13,* 41; image synthesis, 1, 23, 71, 163, 191n23; speech synthesis, 123–24. *See also* latent space

tactical analysis, 155
technical ensembles, 26, 28, 30, 174; and aesthetic diversity, 56–57; and allagmatics, 39; and conversational AI, 115–16; and marginal indeterminacy, 62–63; strategic operations of, 155. *See also* assemblages
technical imagination, 36–37
technical individual, 174–75, 187–88n9
technical object, 28; becoming of, 174; "Black," 93–94; and marginal indeterminacy, 62–63
technicity of elements, 37
techniques, 92–94
TensorFlow, 65, 85, *90*
tensor processing units (TPUs), 59
Terasem Movement Foundation, 110, 194n15
text-to-image models, 1–3, 43, 135, 185n2, 189n6; defined, *45*; and latent space, 62; as likelihood-based generative models, 47
texture detection, 53–55, *54*
theory of mind, 136–37
Thermal Infrared Sensor, 171

Thispersondoesnotexist, 71, 191n24
Thompson, Jeff, 161–65, *162*
thresholds, 28
topology, 12, *13*, 15, 27; nonoptical, intensive, 61; past-future/present-past/present-future, 37; statistico-topologies, 99; of statistics as relational surfaces, 81
Training Humans (Crawford and Paglen), 147–48
transhumanism, 110
"truth," 48, 154
t-SNE (t-distributed stochastic neighbor embedding), 167–68
Turing, Alan, 17, 75, 135–36; imitation game, 122–23
Turing test, 120
Turk, Matthew, 97, 98
typological schemas, 89, 91–93, 108

universality, 177–79; versus generality, 135–36
unlearning. *See* allagmatics; machine unlearning (MUL)

value, data as, 159, 163
variability: drifts, 115, 143–44; eugenic project of constraining, 91, 93, 100, 108; and generalization, 141–42; "normal distribution curve," 100; opening to multiplicity of intelligences, 109–13; quantizing of, 101–2; reduction of, *31, 35*
vectorization, 15, 18, 22–23, 46, 60–62, 97; continuous and discontinuous, 69;

and eigenfaces, *96*; racialized, 91–92, 98–105, 190n18; vector states, *33, 34*
visual cultures, 53–54, 57, 86, 165, 191n19
visual developer tools, platform independent, 85
visual displays, 80–81
Visual Genome, 143
visualization, *14*, 57, 62, 78–79; computer vision models, 8, 156; diagrams, 81; of high dimensionality, 167–69; of "subspace," 99. *See also* artful techniques; invisuality
visual perception, 9, 43, 50–51, 67, 80, 189n4
voice recognition programs, 123–24, 129
Vygotsky, Lev, 136, 196n10

weights and biases, *22*, 67, 156, 171
Weiss, Kenneth, 102–3
Weiss, Steffen, 18–20, *19*
Weizenbaum, Joseph, *120, 121*, 122, 195n4
Weizman, Eyal, 8, 186n3
Weng, 12
Whitehead, Alfred North, 36, 73; "actual occasion," 74, 125, 139
WikiArt dataset, 69, 72–74, 191n21
WordNet, 144, *145, 149*, 149–50
worlding, 88–89

x, y coordinate systems, 95, 98

Zeilinger, Martin, 155
Zylinska, Joanna, 154